VENOMOUS

VENOMOUS

HOW EARTH'S DEADLIEST
CREATURES MASTERED
BIOCHEMISTRY

CHRISTIE WILCOX

SCIENTIFIC AMERICAN | FARRAR, STRAUS AND GIROUX

NEW YORK

Scientific American / Farrar, Straus and Giroux
18 West 18th Street, New York 10011

An excerpt from *Venomous* originally appeared, in slightly different form, in *Scientific American*.

Library of Congress Cataloging-in-Publication Data
Names: Wilcox, Christie, 1985– , author.
Title: Venomous : how earth's deadliest creatures mastered biochemistry / Christie Wilcox.
Description: First edition. | New York : Scientific American/Farrar, Straus and Giroux, 2016. | Includes bibliographical references and index.
Identifiers: LCCN 2016001951 | ISBN 9780374283377 (hardcover) | ISBN 9780374712211 (ebook)
Subjects: LCSH: Poisonous animals. | Venom.
Classification: LCC QL100 .W55 2016 | DDC 572—dc23
LC record available at https://lccn.loc.gov/2016001951

Designed by Jo Anne Metsch

Our books may be purchased in bulk for promotional, educational, or business use. Please contact your local bookseller or the Macmillan Corporate and Premium Sales Department at 1-800-221-7945, extension 5442, or by e-mail at MacmillanSpecialMarkets@macmillan.com.

www.fsgbooks.com • books.scientificamerican.com
www.twitter.com/fsgbooks • www.facebook.com/fsgbooks

1   3   5   7   9   10   8   6   4   2

For those who care about all the creatures of this world,

whether furry or fanged, slimy or scaly,

majestic or misunderstood

CONTENTS

PREFACE

Before our species forged iron blades or wrote the first words, at a time when people cast aside their nomadic lifestyle in favor of settling down and began laying the foundations for the first civilizations; before the birth of Christ or Buddha, before Pythagorus or Archimedes, the people who lived in what is now Turkey built a temple. Today, it is known as Göbekli Tepe, which is Turkish for "Potbelly Hill," and it is the oldest known religious site on earth. Dozens of large limestone pillars remain where devout believers carefully erected them more than one hundred centuries ago—transported and raised by human hands alone, without the assistance of animals or even wheels. But you will find no angels or demons carved into these holy pillars. Instead, the ancient artists chose to decorate their most sacred temple with what they held most dear: the animals relevant to their daily lives, including venomous snakes, spiders, and scorpions.

There is no doubt that venomous animals share with us a deep, rich, and colorful history. They are ever-present in our existence: fear of many of them is innate in humans, found even in the youngest babes; they are vibrant and terrifying in the myths and legends of our tribes and civilizations; and from the dawn of recorded history, we have woven them into our culture. In a way, this book is my offering to these ancient gods—my ode to their fearsome power and my tribute to their incredible scientific potential.

My fascination with these creatures goes as far back as I can remember. When I was a kid, I lived in Kailua, Hawaii, and recall being transfixed by the blue bubbles of Portuguese man-of-wars that would wash up on the beaches near my home. They were just so pretty, so delicate, and I found myself drawn to them, prodding their translucent blue floats with anything I could get my tiny hands on. My fascination would not be deterred, not even after painful stings warned me of their dangerous nature. My compulsion has only deepened over time. When our family moved to Vermont, my mother nearly fainted when I showed her a snake I caught in the yard. During my freshman year of college, I was entranced by the upside-down jellyfish that we caught for our invertebrate zoology lab, and I spent the entire four-hour period gently flipping the creature to watch it swim or sink to the bottom of the glass dish. I kept at it, even as my fingers stiffened and eventually went numb from the mild venom. To this day I can't walk by a touch tank at an aquarium without pressing a finger against an anemone tentacle to feel it stick to my skin as it desperately tries but fails to penetrate my thick cuticle with its harpoon-like stinging barbs. I can spend hours petting the smooth wings of a stingray. I even decided to study the venomous lionfishes for my Ph.D.—a move that my advisor found very amusing. "We just finished a project on moray eels," he told me, with a mischievous glint in his eyes. "Only three people got bitten. I can't wait to see how your project turns out."

Looking back, I'm glad I chose venoms; my colleagues in this field are some of the most open, charming, and electrifying people on earth (though I may be somewhat biased). In my experience, there are two kinds of venom scientists. There are those you might consider lab rats—the people whose initial interest in venoms stems not from the animals but from the molecular complexities present in their toxic secretions. Glenn King, professor of chemistry and structural biology at the University of Queensland and a leader in the search for potential venom pharmaceuticals, was an NMR (nuclear magnetic resonance) structural biologist by training, and it was only when a colleague asked for help determining the structure of a venom toxin that he was lured into the field. Now he's at the forefront of venom bioprospecting, turning the venoms that harm into compounds that heal. Ken Winkel, the former director of the Australian Venom Research Unit at the University of Melbourne, readily admits that he isn't a "snake guy"—he came to study venoms almost by accident, as a byproduct of his initial interest in medical immunology. Similarly, the University of Utah's Baldomero (Toto) Olivera wanted to study neurons and paralysis—cone snails just happened to be a way to do that.

Then, of course, there are the Bryan Frys of the world. Well, really, there's only one Bryan Fry. The head of the venom evolution lab at the University of Queensland is a unique character. *National Geographic* once described him as "a ridiculously handsome adrenaline junkie"—I think of him as the bad boy of venom scientists. He's not one to let others have all the fun, so he travels the world to catch and milk venomous animals, then uses a suite of modern tools to study the venoms from every angle. For his efforts, he's been bitten by twenty-six venomous snakes, broken twenty-three bones, and felt the painful venoms of three stingrays, two centipedes, and one scorpion. When I further inquired how many insects he'd been envenomated by, he laughed. "Who counts bees? You want me to count every fucking fire ant, too?"

Bryan is blunt, brutally honest, perhaps even offensively straightforward. He's also a brilliant scientist and one of the world's experts on venoms. I've known Bryan for years now; I met him when I was a young graduate student just starting on my study of the venomous lionfishes. While I was in Australia to get an up-close encounter with a platypus, I stopped by his lab at UQ, and we shared a pitcher of beer at the Red Room, the campus pub. I realized that although we'd talked at length about various technical topics, I'd never actually asked him what drove him to study venomous animals.

"It's all I've ever wanted to do," he said. Bryan is quick to admit that his study of venoms stems from his love of the animals. On his website, he confesses that while his research is important medically, "it is all just a grand excuse to keep playing with these magnificent creatures!" When he was four, Bryan announced proudly that he was going to work with venomous snakes for a living—and he meant it. Since then he's branched out: Bryan has studied anemones, centipedes, insects, fish, frogs, lizards, jellyfish, octopuses, salamanders, slow lorises, scorpions, spiders, and even venomous sharks. But while it was the animals that drew him to study venoms, it's the venoms themselves that continue to stir his curiosity. What most intrigues him now about venom is, in his words, "how many different flavors of fucked-up it can make you feel."

I, like Bryan and his kind of venom scientist, got into the study of venoms because of my love for the animals. But the more I learned about the intricate complexities of the chemical cocktails these animals produce, the more intrigued I was by the venoms themselves and the more I was drawn to the most dangerous of species. Even the most optimistic projection would suggest that my fascination with venomous animals is going to be a painful learning experience. But nothing worth anything comes without risk. What these animals have to teach us about ecosystems and how species interact is immeasurable; what we can learn about our own bodies from their potent toxins, invaluable; and what they tell us about the fundamental processes of evolution itself, inestimable. I'll happily risk a few ER visits for the opportunity to get a peek at what these animals have hidden in their genes, and to share their secrets with the world. So far, anyway, I have managed to travel around the globe and get up close and personal with a diversity of venomous animals, and I've walked away unscathed.

Well, there was that one monkey bite, but that only required eight injections of immunoglobulin and four rabies shots. And then there was that sea urchin . . .

VENOMOUS

1

MASTERS OF PHYSIOLOGY

Venoms are not accidents, poisons may be.

—ROGER CARAS

If you decided to create a list of the most improbable animals on the planet, the platypus is an easy first pick. The platypus is so peculiar that even the great naturalist George Shaw, who provided the first scientific description of the animal in 1799, could hardly believe it was real. "A degree of skepticism is not only pardonable, but laudable," he wrote in the tenth volume of his *Naturalist's Miscellany*, "and I ought perhaps to acknowledge that I almost doubt the testimony of my own eyes." It is a sentiment I understand. As I sat staring at a large male platypus at the Lone Pine Koala Sanctuary in Melbourne, Australia, I could hardly believe the creature in front of me was real. Even up close, it looked like some kind of masterful puppet, Jim Henson's greatest feat.

Rebecca Bain, known as Beck, the head mammal keeper and one of the people responsible for Lone Pine's two male platypuses, was kind enough to let me in behind the scenes to indulge my interest in the animal. As Beck wrestled the older male from his nest box, I was surprised by his beaveresque tail, duck-like bill, and ottery feet. But while these traits are all fantastically unthinkable, there is one feature of the platypus that stands out among these oddities. It was the feature that drew me to Australia, the reason I came to see the bizarre creatures in person. Beware the male platypus: of the 5,416 currently recognized species of mammals, he alone possesses a venomous sting, using toxic ankle spurs to fight over females.

We know of twelve venomous mammals; all except for the platypus deliver a venomous bite. There are four species of shrew, three vampire bats, two solenodons (long-snouted, rodent-like burrowing mammals), one mole, the slow loris, and the platypus. There's some evidence that the slow loris may actually be four species of slow lorises, which would bump the total to fifteen, but even so, that's still just three handfuls of venomous mammals.

Of the animal lineages, there are venomous representatives in the phyla Cnidaria, Echinodermata, Annelida, Arthropoda, Mollusca, and Chordata—the phylum that includes humans. Compared with other groups of animals, the mammals boast very few venomous members; the Cnidaria, including jellyfishes, anemones, and corals, are an entire phylum—more than nine thousand species—of venomous animals, though if we want to talk sheer numbers, the venomous arthropods, including spiders, bees and wasps, centipedes, and scorpions, undoubtedly reign supreme. There are venomous snails, venomous worms, and venomous urchins. And that's not even including the rest of the venomous vertebrates in the Chordata. There are venomous fishes, frogs, snakes, and lizards.

The term *venomous* carries with it an explicit set of requirements. Many species are *toxic*: they possess substances that cause a substantial degree of harm in small doses (a toxin). We used to think of the terms *toxic*, *poisonous*, and *venomous* as interchangeable; now modern scientists distinguish between them. Both poisonous and venomous species are indeed toxic, for they produce or store toxins in their tissues. You may have heard that everything is a toxin in the right dose, but that's not quite true. A large enough dose can make something *toxic*, but if it takes a lot to kill you, then a substance isn't a *toxin*. Sure, you can drink enough cans of Coke for it to be fatal, but sodas are not considered toxins because the amount it takes for them to be toxic is huge (you'd have to chug liters at a time). The secretion of the anthrax bacterium, on the other hand, is a toxin because even a teeny bit can be deadly.

We can furthµer classify species that are toxic based on how those toxins arrive in a victim. Any toxin that causes harm through ingestion, inhalation, or absorption is considered a poison. Poisonous species, like dart frogs or pufferfishes, must wait for other species to make a mistake before inflicting their toxins. Some scientists would argue there is a third subcategory of toxic, in addition to poisonous and venomous—the *toxungenous* animals—which are essentially poisonous with purpose: toxungenous animals are equipped with poisons, but they're more impatient. Animals like the poison-squirting cane toads or the spitting cobras actively aim their poisons at offenders when they're annoyed, refusing to wait to be touched or bitten, like other poisonous animals, to transmit their toxins.

To earn the prestigious descriptor of "venomous," an organism must be more than just toxic; it must also have a specific means of delivering its dangerous goods *into* another animal. It has to be proactive about its toxicity. Snakes have fangs. Lionfish have spines. Jellyfish have stinging cells. Male platypuses have spurs.

The venomous spurs on the platypus aren't hard to spot. As Beck described the animals and their care at Lone Pine, I stared at the yellow toothlike points jutting from the hind legs. At about an inch long, they are much larger than I had expected. There's no doubt that any wound created by such impressive spurs would

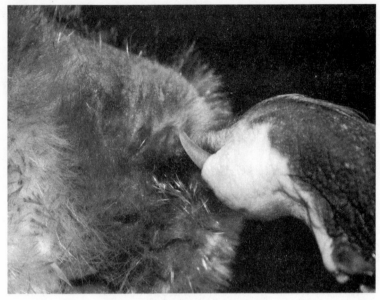

The venomous spur of the platypus
(Photograph by Christie Wilcox)

be terribly painful even without the venom. As I placed my hands within inches of the spurs to get a close-up photograph, I shuddered at the thought of how much it would hurt to be stung by the animal in front of me.

Platypuses are really awfully, terribly venomous. From what I've heard, being stung by a platypus is a life-changing experience, as any deeply traumatic event shapes who you are. Their venom causes excruciating pain for several hours, even days. In one recorded case, a fifty-seven-year-old war veteran was stung in his right hand when he stumbled on what seemed like a wounded or sick platypus while he was out hunting and, concerned for the little guy, picked it up. For his kindness, he was hospitalized for six days in excruciating agony. Over the first half hour of his treatment, doctors administered a total of 30 milligrams of morphine (the standard for patients in pain is usually 1 milligram *per hour*), but it had almost no effect. The veteran said the pain was far worse than the pain from the shrapnel wounds he'd gotten as a soldier. Only when the medics numbed all feeling in his hand with a nerve-blocking agent did he finally feel relief.

Even more bizarre is that the venom the platypus delivers is very different from the venoms of its mammalian relatives. Similar to the animal's outward appearance, with its collection of body parts seemingly taken from other species, it is as if the platypus's venom is composed of a random spattering of proteins stolen from other animals. There are eighty-three different toxin genes expressed in the platypus venom gland, some of whose products closely resemble proteins from spiders, sea stars, anemones, snakes, fish, and lizards, as if someone cut and pasted genes from the entire diversity of venomous life into the platypus's genome. Both externally and internally, the platypus is a testament to the power of convergent evolution, the phenomenon in which similar selective pressures can lead to strikingly similar results in very different lineages. Yet they are also wonderfully unique animals, the only ones we know of that use venom primarily for masculine combat rather than for feeding or defense.

Before she placed him back in his nest box, Beck allowed the platypus to release his rage. She pulled out a towel and dangled it behind him. The animal quickly and gleefully grabbed the towel with his hind legs and began writhing vigorously. The fervor with which he envenomated the cloth was adorable and terrifying. I silently thanked the awkward animal for accommodating my presence, however unwillingly. I'm pretty sure he imagined it was my arm and not the towel he clung to.

Unlike the platypus, many species use needle-like teeth to deliver potent toxins from modified salivary glands, the method preferred by the snakes and most of the mammals. The slow loris,

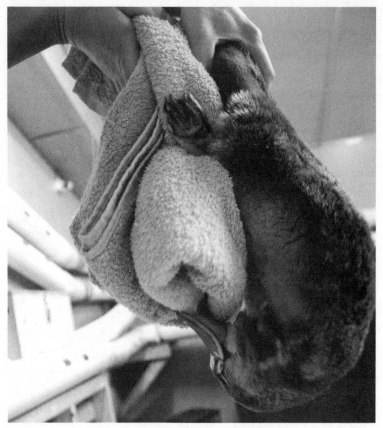

A male platypus angrily envenomating a towel
(Photograph by Christie Wilcox)

however, has its own way of delivering a venomous bite. The small nocturnal primate, a contender with the platypus for most bizarre venomous animal on the planet, uses grooved teeth called tooth combs to deliver its painful venom. But before it can do so, it has to collect venom from glands *on its elbows*. Spiders, centipedes, and many other arthropods also inflict a venomous bite with the aid of fangs or other modified mouthparts. You could even say that some snails deliver a venomous "bite": they strike their food with a harpoon-like structure called a radula, which I think of as a kind of hardened tongue.

Then there are the other stingers. Bees, wasps, ants, and scorpions are the most well-known for their stings, as are the venomous rays (or *sting*rays). A wide array of spiky armaments are employed by caterpillars, urchins, and plants to deliver a potent sting. The Cnidaria possess a unique mechanism—the "stinging cells" (or *cnidocytes*) exclusive to the phylum. They are found along the tentacles of these jellies, corals, and anemones, and can be readily triggered to launch a tube-tethered microscopic needle into whatever comes too close. While we tend to think of them as a venom delivery system, cnidocytes are diverse in form and function, with only some serving to deliver venom. Others discharge glue-like substances or simple hooks to ensnare potential prey.

The two main categories of wounding implements reflect the two main uses for venom: to aid in acquiring or eating prey, or to defend oneself against potential predators. The different uses lead to different selective pressures, and, often, to different venom activities. Those that bite generally use venom predominantly for offense. The stingers, instead, are defensive adaptations. Of course, there are exceptions to each. The scorpion and jellyfish sting to kill prey, and the slow loris bites in defense. And often, species will use their venoms for both, switching from offense to defense as necessary.

Offensive venoms tend to be more physically disastrous. They're often packed with potent neurotoxins to paralyze the intended food or awful cytotoxins that help digest the meal. But they can also be the mildest venoms as far as humans are concerned: if the venoms are intended for an insect or some other species dissimilar to our own, the venom components might not cause the same effects in our tissues as they do in the animal's prey. Or the delivery system may not be tough enough to get through our skin: many species of anemones, for example, are harmless to us because their *nematocysts*—the most common type of *cnida*, the "firing" organ inside each cnidocyte—can't penetrate our dermal layers. Meanwhile, the defensive venoms generally contain different neurotoxins—ones meant to induce horrific, inescapable pain, a warning to select a different dinner. Because they're meant as a warning, most defensive venoms aren't lethal.

One thing is true of all venoms: they're expensive. I don't mean they cost a lot of money on the black market (though some do fetch a pretty penny)—I mean they cost a lot of energy to produce. An animal has to devote hard-earned calories to producing and maintaining its toxic weaponry rather than to other important uses, such as growth or reproduction.

Scientists know that venoms are costly from several kinds of evidence. Perhaps the simplest clue is that even within venomous branches (what scientists refer to as clades) of an evolutionary tree, there are often species that have lost their toxic touch. If venoms are so damned useful evolutionarily, why would any species give up the advantage unless it cost more than it was worth? A shift in diet from active to passive prey, for example, might make a predatory venom far less useful to possess. That's why scientists believe that when the marbled sea snakes switched to eating eggs, they lost their potent venom.

In many venomous groups, there are similar key examples of reduced or lost toxicity. The constrictor snakes could be one such example: some scientists believe that the origins of venomous reptiles date back to before snakes split from their lizard relatives, but the ones that could catch enough prey with constriction had no further need of their venomous bite. The venomous fish lineages are scattered among nonvenomous groups, suggesting that the gain and loss of venom is frequent in fishes. Being toxic just wasn't worth it from an evolutionary perspective.

Something else is true of all venomous animals: we're fascinated by them. Detailed descriptions of venomous animals and
how their bites and stings torment our bodies can be found in some
of the earliest medical texts known, and have been pondered at
length by the likes of Aristotle and Cleopatra. Mithridates VI of
Pontus, a formidable enemy of Rome, was so obsessed with venoms and poisons that he became known as "the poison king." His
father was murdered by poison when he was only twelve, so from
a young age, Mithridates sought a universal cure to any toxin. He
began ingesting small amounts of toxins on a daily basis, believing that he could build an immunity to all poisons over time.

Following on the heels of the poison king were the physicians
Nicander (roughly 185–135 B.C.) and Galen (A.D. 131–201), both
of whom wrote extensively about venomous animals and treatments for injuries inflicted by their diverse toxins. These physicians were considered some of the best authorities on venom and
medicine in general, and well into the fifteenth and sixteenth
centuries, their writings were still read and translated into Latin and
other languages.

Though many physicians and writers would talk about venomous animals, it would take until the seventeenth century for
scientists to begin systematic studies of these dangerous creatures.
Francesco Redi (1621?–1697) was among the first to compile what
was known about snake venoms at the time, and to demonstrate
that they were in fact venoms and not poisons—that many were
harmless if they were ingested, but deadly if injected under the skin.
In the nineteenth century, taxonomy as we know it emerged, and
scientists began classifying and sorting venomous animals.

Strangely enough, though the platypus's spurs were noted in
some of the first specimens—indeed, the first documented sting
was in 1816—scientists would debate for decades whether the
animals were actually venomous. Henri de Blainville (1777–1850),

chair of anatomy and zoology at the University of Paris, created one of the first detailed descriptions of the spur and its associated glands, concluding that the spurs were venom organs, intended to inject toxins "comme cela a lieu dans les serpens venimeux" ("as occurs in the venomous snakes"). Yet in 1823, an anonymous medical commenter assured *The Sydney Gazette* that "I have dissected this animal particularly, to ascertain this much controverted point, and have not been able to trace, either in the *living* or *dead* animal, the virus supposed to be contained in the sac; and I am not *solitary* in my opinion, that *there is no poison*; nor is it, properly speaking, a *gland* which the spur is conjoined to."

"It is my firm conviction that the animal has not the power of instilling poison by its spur," wrote the lawyer Thomas Axford in 1829. He even went so far as to say, "I am so convinced that the spur is harmless, that I should not fear a scratch from one."

The view that the platypus was harmless would hold through the nineteenth century despite reliable reports of envenomations. Even in 1883, the English naturalist Arthur Nicols scoffed at the idea that the platypus was venomous, condescendingly dismissing those wary of the animals: "On seeing me handle my specimen with perfect indifference to the supposed weapon, the black fellow expressed very decided apprehension, and pointed to the spur with gestures of alarm. Here, then, was another example of the ignorance of practical natural history among the Australian natives." The platypus was considered remarkable for its placement as an evolutionary bridge between mammals and reptiles—a mammal that lays eggs!—not because of its potent venom. Scientists were far more focused on its reproduction than its toxins. But, as the nineteenth century came to a close, a growing group of scientists would take interest in venoms, spurring advances in technology that became the foundation of modern studies. The science of venoms was about to take off, just in time to end the debate about the venomousness of the platypus.

The early studies of venoms were largely generated by professionals interested in the clinical implications of dangerous animals. The medical literature is littered with experiments to determine potency, physiological responses, and effectiveness of treatments. These scientists developed and mastered tests for specific activities of different venoms—what we now call functional assays or bioassays. This meant that for the first time, scientists could reliably study the various effects of venoms, often referred to as "activities," such as whether a venom kills cells or stimulates muscle contraction. By comparing the results of such assays with studies in living organisms, researchers were able to gain a better understanding of which venoms attack which systems, and thus develop emergency treatments. They could also begin to compare similar venoms from different species. For example, scientists could determine how effectively the venom from one snake kills red blood cells (a common activity of necrotic venoms) compared to another species in the same genus, allowing them to better understand why certain animals are more dangerous than others.

Scientists had also discovered the secret to treating the worst bites and stings; by 1896, Albert Calmette (a protégé of Louis Pasteur) had created the first antivenom. He was in what is now Vietnam when a flood forced monocled cobras into the village he was staying in, and the sudden increase in bites prompted him to search for a way to treat the deadly envenomations. His solution was to inject a horse with cobra venom, then use its blood serum to treat the bitten, creating the first antivenom. Antivenoms take advantage of the adaptive immune system to create targeted antibodies that bind venom toxins, rendering them harmless. Antivenoms save countless lives every single year, but there is still much room for improvement. Modern scientists are on the hunt for a universal antivenom to streamline treatments and ensure that inaccurate identification of venomous perpetrators doesn't lead to improper dosing. How successful they will be remains unclear.

So while venom science had changed much over the nineteenth century, beliefs about the platypus were still stuck in the past, until the question of the purpose of the platypus's spurs resurfaced in the late 1890s. In 1894, *The British Medical Journal* dared to challenge the conventional view that had prevailed since the 1830s, asking "Is the platypus venomous?" as reliable reports of stings were starting to accumulate. In 1895, the first experiment on a live animal was finally conducted—this one on rabbits—injecting platypus venom milked from the spurs to see how it would act. The results of that test were clear: there is a "remarkable analogy between the venom of Australian snakes and the poison of the Platypus." Chemical analysis of the venom showed that it contained enzymes (known as *proteases*) that could cut proteins, and the study also explained why accounts of envenomation were inconsistent: toxicity was seasonal, with the most toxic venoms produced during the reproductive season, solidifying the argument that the venom was used by males in combat over access to females.

In 1935, the venom scientists Charles Kellaway and D. H. Le Messurier stated unequivocally that the venom was similar to "a very feebly toxic viperine venom." Exactly what was in that venom, though, went unstudied for another thirty years. That's because starting around the 1930s, the study of venoms began to shift away from the almost purely clinical work of early researchers, which had sought to connect venom doses to envenomations in people and how to treat them. While there are still many scientists who investigate venoms and antivenoms from a medical standpoint, a new wave of researchers emerged during the 1940s and '50s who studied venoms by conducting basic research into their molecular mechanisms of action. Technological advances also have allowed researchers to begin to understand the evolution of venoms and their components, which has led to novel insights into their pharmaceutical potential.

One of the problems with studying venoms until very recently was that we didn't have very good ways of teasing apart what was present in the crude substances milked from animals. Chemists had nearly mastered the art of separating vastly different types of compounds, such as lipids and proteins, but such methods didn't finely separate venom components. It was like sorting laundry: they could pull shirts from socks, but couldn't separate based on fabric color, or distinguish long sleeves from short ones. Some venoms have hundreds of different peptides (small proteins), all of which might be soluble in water, for example. That means that an "aqueous fraction," or subsection of a venom separated using water, might contain hundreds of different venom compounds, making it impossible to determine if one or many are responsible for any activity seen when that fraction is injected into a mouse.

Luckily, back in the early twentieth century, the Russian scientist Mikhail Tsvet invented a method to separate pigments from plants that came to be known as chromatography, which, with many later variations and refinements, has helped current scientists to isolate and identify venom components. In chromatography, mixtures are dissolved in a fluid (referred to as the mobile phase) which is then passed through a structure (the stationary phase) with certain properties. This structure can be simply a column of material through which the solution passes, drawn by gravity; or it can have specific chemical properties that make it "sticky" to particular types of molecules. When the mixture is run through the stationary phase, even small variations in the compounds' size, 3-D structure, or chemical properties cause molecules to travel at different speeds, allowing scientists to separate venom components on a much finer scale.

Throughout the 1940s and '50s, new kinds of chromatography were invented, and what is now known as high performance liquid chromatography (HPLC) entered the scene. HPLC, which uses high air pressure instead of gravity to move the solution through a finer-textured column, is now one of the most important techniques in the study of venoms, as it allows scientists to separate venom samples into individual components. And conveniently, during the mid-twentieth century, scientists also invented gel electrophoresis for separating molecules of protein, DNA, or RNA. Gel electrophoresis uses an electric field to pull compounds through a gel by attracting negatively charged molecules to one end, while the gel's properties affect which things will move through it more easily, traveling farther in a given amount of time. You can imagine how much faster a needle can be pushed into molasses than a finger can, for example, if they were both pressed with the same amount of force. When it comes to proteins, electrophoresis is mostly used to separate based on size, giving scientists a rough idea of the number of different proteins present in a venom. It also has become an invaluable method for determining whether genetic extractions or amplifications were successful, and is an absolute requirement in just about every lab that studies venom today.

The modern era of venom research followed on the heels of these two major advances in separation technology. By the 1970s, labs worldwide could examine different components of a venom and their individual activities rather than the crude venom as a whole, and they began separating out the ones responsible for the most noticeable venom actions. Captopril—one of the best-selling drugs of all time, used to treat high blood pressure and heart failure—was isolated (from the venom of a Brazilian pit viper, *Bothrops jararaca*) during this time, as were many other venom compounds.

As a part of his Ph.D. thesis published in 1973, Peter Temple-Smith took advantage of the new battery of techniques to determine the contents and activities in platypus venom. He found at least ten different proteins through electrophoresis and chromatography, and isolated the components that were lethal in mice from ones that caused convulsions. However, the scope of his research was limited, as the separation methods and bioassays still required relatively large amounts of venom (Temple-Smith couldn't complete lethality tests, for example, because he didn't have enough venom to work with). Snakes are easy, as they can be milked repeatedly and produce milliliters, and even liters, of venom fairly readily, but many of the other groups of venomous animals provide only 1/1000th or less of the volume required to run such tests. Though the platypus is capable of delivering upwards of 4 milliliters of venom with each spur, actually getting that much raw material is exceptionally difficult. On average, Temple-Smith and others found they could extract only 100 microliters at a time—too little, in those days, for detailed analyses.

But soon enough, tests miniaturized, and better technologies emerged to determine the shape and structure of different molecules, removing the large-volume requirements that had hindered progress. Scientists who made advances in mass spectrometry (MS) and nuclear magnetic resonance (NMR) won Nobel Prizes in chemistry, and for the first time, these advances allowed them to deduce the chemical composition of larger, more complex compounds like those found in venoms. Even small volumes of crude venom could be evaluated to find compounds that are responsible for key activities such as reducing blood pressure, shutting down nerve impulses, or destroying red blood cells.

In the 1990s, several studies picked up where Temple-Smith left off. Scientists taking a closer look at platypus venom isolated active peptides, two different proteases, and a hyaluronidase (enzymes also referred to as venom "spreading factors" because they cut hyaluronic acid, a major component of skin and the connective "goo" between cells). They could even obtain short sequences from some of these components, and determine that they are similar to snake venom constituents.

Then a new technology completely changed the way in which scientists study venomous animals and their toxins: genomics. Watson, Crick, and Franklin had deduced the structure of DNA in 1953, and thirty years later, scientists invented a method for amplifying fragments of DNA based on their sequence. Polymerase chain reaction (PCR) formed the basis for the first sequencing technology, Sanger sequencing, which is still in use today. The first full gene was sequenced in 1989, and the first full non-viral genome (a bacterium) in 1995. In the twenty or so years since, genetics and genomics have proved to be among the most rapidly changing fields in science. High-throughput technologies can now sequence entire genomes in a matter of hours, and new methods are regularly introduced that produce more information in less time for a lower price. It took years and cost millions to sequence the first human genome, which was finished in 2003— and it's possible that within the next five to ten years, sequencing an entire human genome will cost less than $1,000.

When it came to studying venoms, the genetics revolution opened up avenues that had never been imagined. Scientists could use genes to look at evolutionary relationships, and determine which species were closely related. They could compare the sequences of toxins to other proteins, and begin to understand how venoms evolve. And it wasn't just DNA—scientists have developed methods of sequencing ribonucleic acid (RNA), the step in between DNA and proteins, and can determine which genes are being expressed. Genomics meant that they could sequence every protein expressed in a venom gland to look at the composition of a venom *even without a single drop of it*. Drug companies can build libraries of venom toxins and search them for ones that might act as enzymes, or have the potential to interact with a "target" such as an ion channel (I discuss these later). By combining venom separation and component isolation with genomics, researchers have shifted from the study of venoms to *venomics*. Through such integrated research, we have come to know venomous animals far more intimately than at any point in history, and we have learned that their biochemical prowess is far more impressive than we ever imagined.

Without genomics, we wouldn't be able to compare the dozens of different venom components, and acknowledge how strange it is that a mammal's venom contains toxins that look like those of stonefish, snakes, sea stars, and spiders. We wouldn't know just how bizarre the platypus really is.

Scientists are excited about the potential applications of this burst of platy-gene discovery. "The unusual symptoms of platypus envenomation suggest that platypus venom contains many unique substances which may also be clinically useful," wrote the Sydney-based venom scientist Camilla Whittington and her colleagues. But the platypus still guards some secrets. No one really knows, for example, what part of the venom is responsible for the excruciating pain that accompanies stings. When we learn that, it will tell us even more about the animals, and it may help us understand ourselves better, too. "It is possible," according to Whittington et al., "that this could lead to the discovery of new human pain receptors and thus targets for painkillers."

Back at Lone Pine, after Beck has allowed the shy mammal to return to its home, I stand in front of his aquarium, watching as he swims in search of his shrimpy breakfast. He tumbles and twists, moving through the water with the grace of a fish. When he finds his treasure, he instantly consumes it, his cute little butt waggling side to side with his head as he swallows his meal. The sanctuary won't open for another fifteen minutes, so for now, I have the area to myself. I try to imagine what it must have been like for those early explorers to encounter this weird furball for the first time. If it were me, I would have been entranced. I wouldn't have even considered the notion that platypuses could be dangerous—I would have felt compelled to catch one to get a closer look. Even now, knowing what they're capable of, I feel drawn to the mammal. Only a glass wall keeps me from his serpentine venom. When I meet some of the other infamous venomous animals in the world, there will be no such barrier.

2

DEATH BECOMES THEM

Multiply, vary, let the strongest live and the weakest die.
—CHARLES DARWIN

On July 29, 1997, Angel Yanagihara stepped into the water, as she had so many mornings before. She couldn't count how many times she'd swum the mile course; she was completely comfortable with the route. Besides, she was feeling confident. She had just defended her Ph.D. at the University of Hawai'i at Mānoa, and was looking forward to her first summer as "Doctor." So when a stranger pointed to numerous palm-sized blobs up and down the beach and told Angel not to swim because of the "box jellyfish," she ignored the warning. She had on a Lycra shorty—a thin wetsuit which covered her torso—so she thought she would be okay. And for the first half of the swim, she was.

But as Angel headed back toward shore, she encountered a swarm of box jellies. Her neck, arms, and legs were stung by their thin tentacles. Intense, burning pain coursed through her body as she struggled to swim away from the stinging horde. As she swam, it became harder and harder for her to breathe as fluid filled her lungs, and she was wheezing and gasping for air with every stroke.

"The weirdest part of it was this overwhelming sense of impending doom," she said. She barely made it to a nearby building and rapped on the door with the quarter she kept for emergency calls before losing consciousness.

She came to in an ambulance at the scene surrounded by emergency personnel. The crew treated her stings with vinegar, meat tenderizer, and a hot shower, and suggested she go to the E.R. Angel thought that was overkill, so she signed a release stating that she had gone against medical advice and drove herself home. But the jellies' venom wasn't done with her. "I was bedridden for days in great pain, and none of the approaches I tried brought any relief," she explained to me. Itchy red welts persisted for four months. As a biochemist, Angel was morbidly curious about the toxins that had caused her debilitating pains. But no one knew what made these jellies so devastating—no one had isolated or identified the toxins in Hawaiian box jellyfish. So three weeks later, Angel applied for research funds to find out for herself—and she's been studying jelly venoms ever since.

Box jellies are the deadliest of the members of the phylum Cnidaria, which also includes other jellyfish, corals, and anemones. The cnidarians were one of the earliest lineages of animals; they split from other animals before the development of bones, shells, or brains, more than 600 million years ago. While they lack the systems we consider standard for predators, they are armed with tentacles that contain millions of stinging cells capable of delivering deadly venom in less than a second.

Angel's first major discovery was that the most lethal component in box jelly venom is a pore-forming toxin, or *porin*, which punches holes in the membranes of cells. The porin she discovered—and others, from related species, which have been characterized and sequenced since—will puncture red blood cells, causing them to leak potassium and then hemoglobin before they burst. While the cell rupture (called lysis) seems like the more dramatic effect, it's actually the potassium that makes the jellies lethal. The porins can cause such massive increases in potassium in the blood that they trigger cardiovascular collapse in a matter of minutes. It's an ancient type of toxin, similar to ones found in bacteria. But box jelly venoms also contain a myriad of other components, including snake-like proteins and spidery enzymes.

The National Science Foundation calls the largest of the box jellies—the Australian box jelly, *Chironex fleckeri*—"the most venomous animal on Earth." But what does that really mean? What is the *deadliest* venom on earth? Every venom scientist is asked a variant of this question at some point in their career, and they're loaded questions. Whenever venoms are talked about, they're implicitly ranked by their danger to humans. Pick up a newspaper and read a headline involving a venomous animal, and it doesn't matter what the story is really about—a boy being bitten by a snake on a field trip, the discovery of venomous frogs, whatever—it seems as if the only exciting thing is the lethality factor. There's just something unsettling about the notion that animals that seem so small and delicate can overpower us. A box jellyfish is little more than goo, yet it can kill a man in less than five minutes. A spider or a scorpion can be unceremoniously crushed under our feet, yet some of their venoms can take us out just as easily.

The threat venoms pose is also of the utmost importance evolutionarily. Natural selection occurs when individuals survive and reproduce more than others. So anything that affects survival directly has a marked impact on a species, and has the potential to shape the way in which that species evolves. Because of their deadly nature, venomous animals have intimate relationships with other species, especially those they prey upon. But because venoms are deadly to more than just their prey, venomous animals also shape the evolution of species they don't naturally eat—including us, in many cases. They play pivotal roles in the complex interactions that make up entire ecosystems, and they affect other species on a global scale.

So what the deadliest venom is depends on several factors. The simplest answer is that the deadliest venom is the one injected into you—a lesson Angel learned the hard way when a predawn plunge nearly killed her. Scientifically speaking, there are various ways to measure "deadliness." One of the most common scales is Median Lethal Dosage, referred to as LD_{50}. LD_{50} is the dose of a toxin (often in mg/kg: milligrams of toxin per kilogram of body weight) that it takes to kill half of a number of test animals (rats or mice are most common, but scientists have used everything from cockroaches to monkeys to test various forms of venom).

An LD_{50} value is a crude measurement of potency; a substance with a low value is considered to be extremely toxic, as it is likely to kill even in low doses. Water has an LD_{50} of >90,000 milligrams per kilogram (mg/kg), which is generally considered harmless, but it can be deadly if you drink more than six liters of water at once (I don't recommend trying it). The LD_{50} of botulinum toxin, on the other hand, is estimated at 1 *nanogram* per kilogram—a nanogram is one *millionth* of a milligram—making it one of the deadliest substances known to man. Sixty nanograms, or 60 *billionths* of a gram, is enough to kill the average person. A handful is enough to kill everyone on the planet if divided equally among them. But use the compound in minuscule amounts, like a tenth of a nanogram injected into the forehead, and it is a favorite among celebrities and those overly concerned with wrinkles (its pharmaceutical name is Botox).

The trouble with using LD_{50} values as your only measure of "deadliest," however, is that LD_{50} can vary depending on both the method of exposure (injecting the venom into the veins rather

Phylum	Animal Lineage	Representative Species	Scientific Name	LD_{50} mg/kg (route)
Cnidaria	Jellies and Anemones	Australian box jelly	*Chironex fleckeri*	0.011 (i.v.)
Arthropoda	Spiders	Black widow	*Latrodectus mactans*	0.90 (s.c.)
	Scorpions	Fattail scorpion	*Androctonus crassicauda*	0.08 (i.v.)–0.40 (s.c.)
	Centipedes	(no common name)	*Otostigmus scabricauda*	0.6 (i.v.)
	Moths and Butterflies	(no common name)	*Lonomia obliqua*	9.5 (i.v.)
	Bees, Wasps, and Ants	Maricopa harvester ant	*Pogonomyrmex maricopa*	0.10 (i.p.)–0.12 (i.v.)
Annelida	Worms	Bearded fireworm	*Hermodice carunculata*	*Not yet determined*
Mollusca	Snails	Geographer's cone	*Conus geographus*	0.001–0.03 (estimated in humans)
	Octopus and Squid	Blue-ringed octopus	*Hapalochlaena* spp.	*Not yet determined*
Echinodermata	Urchins	Collector urchin	*Tripneustes gratilla*	0.05 (i.p.)–0.15 (i.v.)
Chordata	Stingrays	Round stingray	*Urolophus halleri*	28.0 (i.v.)
	Fishes	Stonefish	*Synanceia horrida*	0.02 (i.p.)–0.3 (i.v.)
	Amphibians	Bruno's casque-headed frog	*Aparasphenodon brunoi*	0.16 (i.p.)–>1.6 (s.c.)
	Elapids (Snakes)	Inland taipan	*Oxyuranus microlepidotus*	0.025 (s.c.)
		Coastal taipan	*Oxyuranus scutellatus*	0.013 (i.v.)–0.11 (s.c.)
	Vipers (Snakes)	Mohave rattlesnake	*Crotalus scutulatus*	0.03 (i.v.)
	Mammals	Short-tailed shrew	*Blarina brevicauda*	13.5–21.8 (i.p.)

A table of the most lethal LD_{50} values for members of venomous animal groups; routes of administration—in mice unless otherwise noted—are listed after each value in parentheses (s.c.=subcutaneous, i.p.=intraperitoneal, and i.v.=intravenous).

than the flesh of your subject, for example) and the species used as the guinea pigs. The values in the table on page 25 are all in mice, but even then, the route of administration matters. If a scientist injects the venom directly into the veins of a mouse instead of under the skin, the coastal taipan is the deadliest venomous snake, with an LD_{50} of 0.013. But if subcutaneous LD_{50} is used instead, the same snake falls several slots, with an LD_{50} of 0.099—almost a *tenfold* difference in potency. Furthermore, we haven't yet calculated all values for all species. We don't know if the intravenous LD_{50} for the inland taipan would beat out its close relative, simply because no one has figured it out yet—we've only tested its venom subcutaneously.

LD_{50} values require careful extraction of the venom and laboratory study of its effects, and while scientists have done a lot of work on many venomous species, there are ones missing from the scientific literature that might be among the most potent in the world. We know that *tetrodotoxin*, the main component of blue-ringed octopus venom, has a subcutaneous LD_{50} of 0.0125 mg/kg—but no one has tested the crude venom as a whole. The porin toxins of the box jellies that stung Angel—the Hawaiian box jelly, *Alatina alata*—have LD_{50}s ranging from 0.005 to 0.025 mg/kg, but no one knows (yet) how much of this toxin is delivered by a stinging tentacle. Similarly, some zoanthids (a relative of corals) possess *palytoxin*, which, with an LD_{50} of 0.00015 mg/kg, is one of the most toxic substances on the planet, but to what extent this toxin appears in their venom rather than throughout their tissues (as a poison for predators) has yet to be determined. And no one has even begun to calculate the LD_{50} for the venoms of many species. The venom of the flower urchin is probably one of the deadliest on earth, as this is the only human-killing urchin we know of, and its much weaker cousin, the collector urchin, has a potent intraperitoneal LD_{50} of 0.05 mg/kg, but no one has done the experiments. The dreaded Irukandji jellyfishes (often less than an inch long!), a type of box jelly, induce Irukandji syndrome, which can escalate to the point of death by cerebral hemorrhage. But until we can determine exactly which species cause this syndrome (there are at least sixteen jellies identified, so far, that are culprits) and collect enough of their venoms to run a lethal dose experiment (not easy given that individuals from some of the species easily fit on your thumb), we'll never know their true potency.

Because LD_{50} values are calculated using mice and rats, they don't necessarily tell you much about how dangerous something is to people. Species can react very differently to venoms; for example, guinea pigs are ten times more sensitive to black widow spider venom than mice, and two *thousand* times more sensitive than frogs. Just because an animal's venom has a low LD_{50} in rats doesn't mean you'll definitely die if you get bitten or stung, and a high one doesn't mean you're safe. Perhaps a better way to look at deadliness is to compare case-specific mortality: the percentage of people who don't survive. The Australian box jelly, for example, kills less than 0.5 percent of the people who are stung every year, and even the dreaded inland taipan has become effectively nonlethal since an antivenom was introduced in 1956 (though prior to that it killed people at a rate of almost 100 percent).

The common krait and king cobra are much more efficient killers. The venom from these large snakes (delivered by short, fixed fangs, as opposed to the long, folding fangs of viperids) isn't all that deadly by the drop, but what they lack in potency they make up for in volume. King cobras can deliver up to 7 milliliters of venom with every bite—enough to kill twenty people! Scientists estimate that anywhere from 50 to 60 percent of king cobra bites are fatal, compared to about 2 percent of venomous snakebites overall. Bites by the common krait are also known for their high fatality rate, which is anywhere from 60 to 80 percent.

Bites from these nocturnal snakes tend to be relatively painless, so victims are lured into a false feeling of safety. It isn't until hours later, as progressive paralysis slowly suffocates them, that they realize they should have sought medical care—and krait antivenom—immediately.

Only a few venomous animals are known to kill at the levels of snakes. The *Lonomia* moth caterpillars stand out with their 2.5 percent mortality rate in humans (prior to the introduction of *Lonomia* antivenom in the mid-1990s, the seemingly innocuous caterpillars had a case-fatality rate of 20 percent). But when it comes to lethality rates, it's the deadliest cone snail in the world, *Conus geographus*, which takes the prize for the invertebrates, with a case-fatality rate of 70 percent. The incredibly high rate reflects the speed with which it kills—those who die succumb in a matter of minutes from sweeping paralysis.

While mortality rates more accurately reflect how dangerous these venoms are to us, they don't tell the whole story. The snakes currently rank pretty low on the lethality death scale (with a few exceptions) because we developed antivenoms to treat their toxic bites. And measuring deadliness by mortality is confounded by the fact that mortality rates vary widely based solely on access to medical care, and they don't tell you how likely you are to get killed because they don't tell you anything about frequency. For example, if I get harpooned by a cone snail radula, that 70 percent mortality rate tells me I should start to worry: but what are the odds that such an event will happen?

The most ecologically and evolutionarily relevant way to measure deadliness is by examining the total number of deaths caused by venomous animals every year. In that way, you can calculate per capita risk. Even today, snakes are one of the top venomous killers worldwide. The Big Four snakes in India—the Russell's viper, the saw-scaled viper, the Indian or spectacled cobra, and the common krait—kill tens of thousands of people every year even though their venoms aren't particularly potent by acute measures such as LD_{50}. Their venoms are 30 to 110 times less potent than those of some of the other elapid snakes, and the mortality rates from the bites of two of them are low, but the Big Four are found in and around major population centers, where they frequently come into contact with people, leading to more bites. While there is an antivenom available for the four species, many bites occur in poor communities where medical access is limited, so bites that are perfectly treatable can be fatal. Similarly, there are tens of thousands of deadly snakebites in sub-Saharan Africa, even though the snakes responsible aren't particularly "deadly" by any other measure than total number of kills. The snakes with the deadliest venoms by LD_{50} values or mortality rates live in remote areas, tend to isolate themselves from locations where humans are common, and rarely end up biting people—so they are rarely responsible for human deaths.

Then, there are some that kill more people than they naturally would, aided by human accomplices. Despite the dangers to the perpetrator, there is a long history of using venomous animals for violence—so much that in several ancient cultures, specific punishments are detailed for those who commit such crimes. For example, in the ancient Hindu code of laws known as the Gentoo Code, it is written that "If a man, by violence, throws into another person's house a snake, or any other animal of that kind, whose bite or sting is mortal" then he shall face a hefty fine and, most important, be forced to "throw away the snake with his own hand." Then there are the Vish Kanya—legendary young women assassins in India during the Mauryan Empire (321–185 B.C.) who were said to be bitten by snakes from birth until their very blood and saliva were so toxic with venom that they could kill with a kiss.

Even in modern times, we've sought to get away with murder by blaming venomous animals. Thomas Burton's book *The Serpent and the Spirit* tells the story of Glenn Summerford, a snake-handling preacher who had supposedly left a history of violence and substance abuse behind in 1982 when he abruptly surrendered himself to the Lord. By 1991, he was a pastor of the Church of Jesus Christ with Signs Following, in Scottsboro, Alabama, and was considered holy by his congregation because he would drape large rattlesnakes over his shoulders and place small ones in his pockets, all the while believing that his prayers were enough to keep him safe from their potent venoms. His wife, Darlene, was also a snake handler; she kept pictures in her purse of her favorite snakes like a proud mother does of her kids. The couple kept more than a dozen live venomous snakes, including western diamondbacks and canebrake rattlers, in a shed by their house. But while Glenn appeared as a man of God to his followers, his inner demons never fully disappeared. In February 1992, he would find himself on trial, charged with attempted murder by snake.

Venomous murder is, on the surface, a perfect crime. After all, so long as you can make it seem like an accidental attack, no one would assume foul play. Venomous animals are well-known for the deaths they accidentally cause; all one has to do is find a way to get a deadly snake, spider, or scorpion to bite or sting the intended victim, and no one will be the wiser. There is the small problem, though, of procuring and keeping a lethal beast until the crime can be committed. For Glenn, that wasn't an issue. He had snakes at his disposal.

According to Darlene, in October 1991, Glenn got drunk. Very, *very* drunk. He grabbed Darlene by the hair, put a gun to her head, and dragged her to the snake shed, where he forced her to put her hand in one of the snake boxes containing a western diamondback. Naturally, it bit her. As her hand swelled, he made her run errands, of all things—returning movies to the video store, then a quick trip to the liquor store—then drove her home again. He dragged her back out to the shed and made her reach for an angered canebrake rattlesnake. She lay on the couch slipping in and out of consciousness as the venom coursed through her veins. He forced her to write two suicide notes. Then he passed out drunk. As he slept, she crawled to the kitchen and called her sister for help. Darlene spent several days in the hospital, and she survived.

Glenn, of course, told a very different story. "I believe she went to get a snake and was goin' to kill me with it while I's asleep," he said. According to Glenn, his wife was unfaithful and sought a way out of their marriage without revealing what he believed were her infidelities in divorce proceedings, so while he slept, she went to the shed and tried to grab a snake to kill *him*—and was bitten in the process. "But I ain't got no proof of it." The jury sided with Darlene, and Glenn was sentenced to ninety-nine years in prison for the attempted murder of his wife. No one seems to know what happened to the snakes.

Other infamous cases have involved other failed attempts to kill by snakebite. In 1942, Robert "Rattlesnake" James made history as the last man executed by hanging in the state of California. He was convicted of murdering his third wife after a life insurance investigation revealed foul play. James tried to make the death look accidental, and purchased rattlesnakes through a friend, forcing them to bite his soon-to-be-ex-wife's leg. But when the venom hadn't killed her several hours later, he impatiently drowned her in the bathtub instead and left her body in their fishpond, hoping it would look like an accident.

Of course, if you don't want to risk being bitten yourself, you could always hire someone more familiar with snakes to kill someone *for* you. In a case from India, a man paid a kidnapper and a snake charmer to kill his elderly parents over a property dispute. According to the police, the couple and their driver were kidnapped, and while the driver was in another car, a snake charmer sitting in the passenger seat coaxed an Indian cobra to bite the intended victims. They each collapsed immediately, and the driver was told to take them to the hospital and report that they were bitten by a snake. Both died within an hour of their bites from the venom's effects, but the son and his accomplices were later caught and punished for their crimes, as Indian law holds all parties equally accountable in the case of contract killing.

Intentional death by snakebite isn't always a murderous act. The Egyptian queen Cleopatra was said to have committed suicide by allowing an asp—most likely an Egyptian cobra—to bite her arm. That's the story favored by the Roman ruler Octavian (later known as Augustus), who paraded an effigy of her body with a snake attached to it through the streets to announce his triumph over her forces in war. Such a death would have been viewed as noble and befitting a great queen, for in Egypt, death by snakebite was thought to give a person spiritual immortality. In Alexandria, snakebite was generally considered a humane method of execution. But in many other cultures, including China and India, it was deemed a death for the most heinous of criminals. Europeans reserved snake pits for those who they felt deserved the cruelest of deaths, including the infamous Viking conqueror Ragnar Lothbrok, who was responsible for countless deaths from raiding and plundering the English countryside.

Luckily, we can find evidence of proposed treatments for snakebite in the earliest known medical texts, but even today, with our antivenoms and modern medical technology, as many as 100,000 people are killed worldwide by snakebites every year. Their high death toll is why we're so afraid of these fork-tongued reptiles, but that fear was once strongly linked to respect. The earliest of cultures revered snakes; even the serpent in Eden was a symbol of knowledge as well as evil, and in many Asian traditions, serpents are associated with intelligence and elegance.

From an evolutionary standpoint, reverence is probably due: so much death means snakes have had a lot of evolutionary influence on us. And if you ask Lynne Isbell, professor of anthropology at the University of California, Davis, that influence has been profound, as she has written that snakes may just be the driving force behind our big brains.

Scientists still argue about how and why we evolved into the bipedal, relatively hairless, intelligent beings we are today. In particular, there are many hypotheses that seek to explain why our brains became so big compared to those of other mammals. One of the most accepted ideas is that primates as a whole developed bigger brains as a byproduct of the demand for more-acute vision to guide reaching and grasping, certainly an important ability for a group of animals that seek a life high in the trees. Better hand-eye coordination was necessary for locomotion, but also to exploit new foods—flowers and fruits—that the ancestors of modern plants had just begun producing. The importance placed on vision is key: big brains didn't evolve for thought or reason, but rather to swiftly process visual information. Other mammalian groups honed senses of smell or hearing, but the primates focused instead on their sight.

Lynne Isbell was the first anthropologist to propose that while primates' arboreal lifestyle and predilection for fruit may have contributed to their visual systems, the real driver of acute vision was predation pressure. It wasn't just that snakes were around— they had been feeding on mammals for millions of years and, as Isbell believes, helped drive better vision in many mammalian groups. Instead, the Snake Detection Theory, as Isbell refers to it, posits that as our ancestors split from the lemurs and other early primates, they were forced to adapt to a change in the predators they knew all too well. In Asia or Africa around 60 million years ago, the snakes became more venomous (though scientists aren't quite sure why then and there). The Viperidae and Elapidae were born.

Armed with a more advanced venom delivery system, the snakes could deliver toxins with deadly efficiency. Almost all the species that can kill a human being are found in those two venomous families. The long-fanged Viperidae, generally referred to as vipers, include the well-known rattlesnakes and other infamous pit vipers, as well as Old World "true" vipers, while the Elapidae, or elapids, boast the most potent snakes on the planet, such as the inland taipan and the black mamba. When these families emerged, they changed the relationship between primates and snakes. The slithering beasts had become an even greater threat. Snakes changed the way they hunted, too, remaining motionless until the last moment. Primates that could detect these often cryptic predators survived and reproduced—detection that required acute 3-D vision and the ability to spot well-camouflaged shapes.

While all primates are good snake spotters, the ones our lineage emerged from—the catarrhine primates—are better than others. This makes sense, because other primates made it to the New World before the venomous snakes did, relaxing the pressure for enhanced vision in the South American line of our cousins, the platyrrhines. It would take anywhere from 12 million to 32 million years for the deadly snakes to catch up with the platyrrhines. As expected if Isbell's hypothesis is right, there is a greater range of variation in visual systems in these New World monkeys, for their vision has been allowed to fluctuate without the intense pressure from snakes. The primates that stayed behind in the Old World (including the apes) had no such respite, and their vision has become remarkably consistent—and remarkably good at detecting snakes.

All primates, including humans, are innately afraid of snakes. We're also very good at spotting them. We can detect snakes that are well hidden in a cluttered environment and snakes at the edges of our visual field. We spot snakes in conditions where we can't spot spiders or other potentially dangerous animals. We can see snakes before we realize that we see them, a phenomenon referred to as preconscious detection. We respond physiologically with anxiety when images of snakes are flashed in front of our eyes on a computer screen at speeds so fast that we do not actually

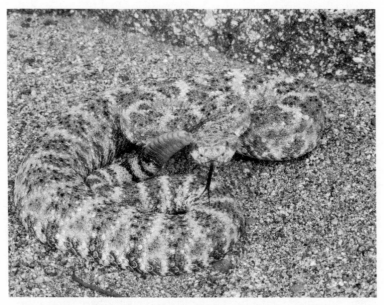

Our eyes have no trouble spotting the speckled rattlesnake,
even with its impressive camouflage.
(Photograph by Chip Cochran)

register seeing them, but we have no such reaction to nonthreat-ening shapes like mushrooms or flowers. All of which suggests that our eyes and our visual neurology are adapted to avoid snakes.

The fruit-eating and tree-dwelling habits of our ancestors might have driven the primate lineage toward improved vision over other senses for predator detection, such as smell, but according to the Snake Detection Theory, it was the deadly snakes that main-tained the evolutionary pressure. To allow for such neural com-plexity as this vision required, our brains had to grow. In addition, our high-sugar diet gave us the energy needed for cranial enlarge-ment. And, thanks to the venomous snakes, the stage was set for our lineage to continue to evolve bigger brains. Isbell hypothesizes that the switch to bipedalism gave the final nudge our ancestors needed—with hands free, they were able to combine gesture and vocalization more effectively, eventually leading to the beginnings of language, which allowed for even greater social complexity and propelled even more changes to our big, human brains. The rest, as they say, is history.

Our relationship with snakes has continued through the ages, and they remain one of our deadliest venomous threats. But snakes *aren't* the deadliest venomous animal when it comes to total kill rate per year. The group that wins by this measure is a *really* dark horse in the race for the deadliest venoms. Perhaps it's the arach-nids, like the black widow spiders? Not even close. What about the humble honeybee? It's true that bees and their relatives kill more people every year in the United States than snakes, scorpions, and spiders *combined*—ten times as many, in fact. Most of us wouldn't have considered the family Hymenoptera—bees, wasps, and ants—as competitors for the dubious trophy of deadliest venomous animals on the planet, but they are certainly vying for the title based on annual kill rate. Of course, their venoms aren't killing people because of low LD_{50}s or even high mortality rates—rather, people are stung with extremely high frequency, and some proteins in insect venoms are potent allergens, so people routinely die from anaphylactic shock after they are stung. But the Hymenoptera are still not #1. The venomous group responsible for *hundreds of thousands* of deaths every single year—orders of magnitude more than any other group—killing more people than even other people do? The Culicidae, better known as mosquitoes.

Mosquitoes use their venom to gain easy access to our blood. They produce vasodilators (compounds that open our capillaries to speed blood flow), anticlotting and antiplatelet agents to ensure the wound stays open while they feed, and anti-inflammatory compounds that prevent our immune system from signaling their invasive presence. Their venoms are well adapted for their hematophagous (or blood-feeding) lifestyle, and often, we don't even notice that we've been envenomated until long after the mosquito has had her fill. The venom is fairly harmless on the acute toxicity scale. Similarly, the mortality rate from mosquito bite is incredibly low: I've received hundreds of bites in my lifetime and I'm still alive and kicking. Few people are truly allergic to their bites, so anaphylaxis isn't to blame for their kills, either. Instead, mosquito venom is deadly not because of the venom itself, but because of what hides within it: mosquitoes are the vector for a suite of infectious diseases including malaria, dengue, and yellow fever.

Although their deadliness is almost entirely attributable to other species that hitch a ride inside their bodies, if mosquitoes were not venomous, they wouldn't make such perfect disease vectors. Venom is what allows them unfettered access to our circulatory system, and it is through the act of envenomation that the diseases carried by mosquitoes find their way into their unfortunate hosts.

So their venomous nature is directly to blame for the death toll from the diseases they transmit.

And what a death toll it is. Malaria claims more than 600,000 lives every year. Add to that 30,000 from yellow fever, 12,000 from dengue, and 20,000 from Japanese encephalitis. Then there's chikungunya, West Nile, Rift Valley fever, and other encephalitises. We have to include the 40 million people disfigured by lymphatic filariasis; though elephantiasis isn't as fatal, it has left many unable to live normal lives. And new mosquito-borne disease threats, like Zika, keep emerging. The annual death toll is high enough that you may wonder why we don't just wipe the damned things off the planet altogether.

Indeed, the esteemed journal *Nature* asked scientists what would happen if mosquitoes were forced into extinction. Some felt that little would change. However, others noted that the loss of mosquitoes would invariably put some predation pressure on other insects, and might have unforeseen effects. Mosquito larvae make up a large amount of biomass in aquatic systems, and are a component of a functioning wetland ecosystem. Though they may not be the only source of food for most species, the loss of such a large amount of biomass would certainly be felt by fish, frogs, and bats that feed on mosquitoes regularly. Similarly, the plants that mosquitoes pollinate would be hindered by their loss, if not pushed to extinction as well.

Perhaps the largest impact would be the lack of their feeding behavior. In the arctic, mosquito populations can be so dense that caribou herds will alter their migration course just to avoid them. The bloodsuckers have been known to drain 300 milliliters of blood—almost a soda can's worth—from every single caribou in a herd *daily*, so it's no wonder the caribou go to such great lengths to avoid them. Even small alterations by such a large herd have a dramatic effect on the land they trample, for thousands of caribou travel in these massive herds. The arctic would feel the effects of lost mosquitoes—so, too, would everyplace else.

In some cases we might applaud the changes. After all, avian malaria has proved as much a problem for birds as human malaria has for us. Mosquitoes were introduced in Hawaii, for example, and until recently, the birds there had had no reason to fear the parasitic scourge. But with the mosquitoes came the disease, and now, native Hawaiian species are being wiped out at elevations where mosquitoes flourish. Mosquitoes can't survive at high elevations due to cold temperatures, so the birds that live at upper altitudes are free from the biting pests. Only the highest peaks on Maui and the island of Hawaii provide refuge.

But to think we could remove twenty-five hundred mosquito species from this planet without serious consequences is evidence of either our cognitive limitations or our unflappable hubris. Mosquitoes have been on this planet for a hundred million years, co-evolving intimate relationships with countless species—including our own. They've kept human populations under a certain level of control, and they've affected us at the genetic level, promoting mutations such as sickle cell to persist in populations despite their otherwise negative effects. And if we were to eradicate mosquitoes tomorrow, the effects on our species alone would be huge. Removing mosquitoes would be like throwing a boulder into a pond; there would be a huge splash at the center, but the consequences would ripple outward.

Besides, we've only begun to understand the chemical ingenuity of these bloodsuckers. Mosquitoes have some of the simplest venoms on the planet, with only a few dozen major venom compounds, yet we still don't know what many of the components do. With the newest wave of drug discovery focusing on venoms, it would be prudent for us to understand the molecules created by these biochemical engineers before we destroy them.

————————

For better or worse, the lethal nature of venoms has inspired both science and superstition, and has drawn the attention of some of the greatest minds in history. If venoms weren't so deadly, we wouldn't have spent so much time and effort as a species studying the animals that manufacture them or the ways they manipulate our bodies' most vital systems, and we wouldn't know just how complex and incredible these toxic substances really are. And if venomous animals weren't so deadly, they wouldn't play such ecologically important roles in ecosystems around the globe.

Whether venomous snakes drove the evolution of human minds or not, there's no doubt that their deadly nature has influenced our evolution and continues to do so even now, though we're no longer on the menu. We know that other venomous species—like the mosquitoes—have shaped our evolutionary past, provoking changes in humans at the genetic level. Whether we like it or not, venomous animals are evolutionary "enemies" that have helped make us who we are today. And they're a factor in our future as well. Our evolutionary fate, like those of some other species we're about to meet, is forever intertwined with the Darwinian destinies of the snakes, jellies, and other venomous organisms.

3

OF MONGEESE AND MEN

Every possible thought of horror filtered through my mind and in one wild insane moment, I thawed the venom, drew up 1 cc into an insulin syringe and cleaned off a spot on my left arm. I held my breath, stuck the needle in and pushed the plunger.

—JOEL LA ROCQUE

When you think about it, your average snake isn't very scary. Its skin is thin and fragile and hardly protects its flesh. Its bones are brittle and frail, easy to crush, even with the smallest of jaws. Really, if it weren't for the threat of potent toxins, snakes would be easy prey for a wide variety of species. The biggest boas or anacondas might have enough of a size advantage to deter smaller predators, but most snakes are small and defenseless without their fang-delivered toxins—which is why nonvenomous snakes often rely on speed or camouflage to stay off the menu. Some even fake the look of more-venomous relatives and hope that the ruse works on their worst enemies. In fact, snakes in general have benefited from the innate fear branded into our genes, and those of many species, by the venomous minority. The reason the deadliest snakes are the most terrifying—to us and to their predator and prey species—is that they have potent venoms, and they know it. And they will happily signal to any animal that comes too close that they are not to be messed with. Rattlesnakes rattle. Puff adders puff, inflating their bodies with air to appear even larger, hissing loudly. Cobras extend their magnificent hoods and stare confidently at anyone who gets too close.

Yet there are some animals that simply don't care if venomous snakes are armed with toxins. They eat them anyway. Everywhere that venomous snakes are found, they coexist with at least one species that seems to have no trouble turning some of the planet's most potent predators into palatable prey. The term for such species is *ophiophagous*, or, quite literally, snake-eating. Other than their dietary preferences, there are no obvious traits which connect these disparate lineages. They don't act differently from their relatives in any way that would give them an advantage against venomous snakes. They don't have armored skin or some other kind of morphological adaptation which allows them to feed on such dangerous delicacies. None of them, it turns out, feed exclusively on snakes—they have diets which consist of many species, including reptiles, small mammals, and sometimes birds. But what they do share is an uncanny ability to shrug off even the most potent snake venoms. A dose that would easily kill a man barely makes an impact on these ferocious little creatures, and their tendency to approach cobras and the like without fear has given them some of the fiercest reputations on the planet. For example, in Rudyard Kipling's *The Jungle Book*, a young mongoose named Rikki-tikki-tavi saves a British couple living in India from cobras—an act often described as "valiant" or "brave." But honey badgers, as well as several opossums, at least two hedgehogs, the aptly named snake eagles, some skunks, and the "heroic" mongooses aren't brave, for to be brave, there has to be

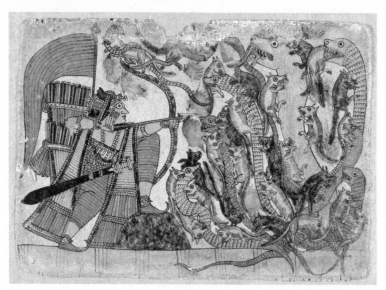

"Fight of the Mongoose and the Serpent Armies," a watercolor depiction
of the age-old battle between snake eaters and their prey

real danger that an animal confronts. Rather, these animals have
nothing to fear, since they have developed molecular mechanisms
that render snake venoms harmless.

Just as venomous animals have toxins that have evolved over
millions of years to dismantle the most precious systems of their
prey, there are species that have evolved defenses against even the
most debilitating venom toxins. Such impressive adaptive traits
arise through *coevolution*, which can be thought of as two (or
more) species having a marked mutual impact on each other. Due
to a high degree of interaction—in this case, a predator-prey
relationship—species become ecologically interdependent. If the
antelope gets faster, for example, the cheetah has to up its speed
or go hungry. When one species changes, the other faces a strong
selective pressure to change, too, or face extinction. The potent
venoms snakes employ don't just work on their prey; they are an

excellent defensive deterrent. So would-be snake eaters that, through random chance, had even the slightest bit of natural resistance to venoms were able to benefit from continuing to eat snakes. The added food resource allowed them to survive and reproduce when their brethren could not. Over time, that resistance became stronger and stronger. But how do the animals do it? Scientists are beginning to understand the molecular adaptations that protect these animals, and to investigate whether other species, including humans, can become immune to the toxins as well.

Part of the mystery has long since been solved: scientists know that all mammalian immune systems have some ability to respond to venoms; the only question is whether their response is fast and furious enough to ensure survival. Mammalian immune systems consist of two equally important responses to anything that enters the body which doesn't belong: *innate* and *adaptive*. Our innate immune system is responsible for the first wave of defense when confronted with an invader, whether it's bacteria, a virus, or venom. The adaptive immune system, on the other hand, "remembers" previous attackers, and allows us to mount a better response the next time those same invaders attempt to enter the body.

Mammalian bodies have many ways to prevent foreign particles from ending up in the wrong place. Our skin and the mucous layers of our nose, throat, and gut act as physical barriers, but they also constantly pump out antimicrobial compounds. As soon as invaders are detected beyond those initial barriers, the innate immune system jumps into action. Injured or infected cells release compounds that tell nearby immune cells that they're in trouble. Mast cells, which are packed with histamine (a vasodilator, which opens blood vessels to stimulate flow) and heparin (an anticoagulant), and macrophages, which are the body's cellular army, engulfing anything that doesn't belong there, start the assault and trigger inflammation. Inflammation, which is marked by redness, heat, swelling, and pain, may be an uncomfortable side effect—but it's a carefully regulated response that the body uses to kill certain types of bacteria and viruses without doing too much damage to itself.

Macrophages go on a rampage, "eating" bacteria, viruses, and other foreign particles in a process called phagocytosis. Once the offending material is inside the macrophage, it is ripped to shreds by enzymes that are stored in special compartments. If the macrophages don't destroy all the invading forces quickly, they release compounds that attract neutrophils, a kind of white blood cell, which also engulf and destroy. In addition to joining in on phagocytosis, neutrophils ramp up local inflammation, release toxins, and set traps made of DNA.

As the battle continues between the innate immune system and the unwelcome visitor, the adaptive immune system kicks in. Dendritic cells present at the site of inflammation engulf whole bacteria, viruses, or proteins and other particles and, like macrophages, chop them to bits. But they don't stick around to fight—they take their precious cargo and head to the nearest lymph node. They take the bits and pieces they collected and present them on the outside of their cell membranes to T cells, using proteins called major histocompatibility complex, or MHC, molecules. All T cells are born with specialized receptors, and only those whose receptors match what the dendritic cell is presenting will be activated. When this occurs, the T cell divides like mad. Some of the clones stick around as *memory* T cells (and will be ready should the same attackers surface again in the future), while others turn on killer T cells, which go help the macrophages and neutrophils. A third subset activate B cells, which are the body's antibody factories. These B cells, like the T cells, are preconditioned to have certain receptors, so not all B cells will react to a given T cell. But

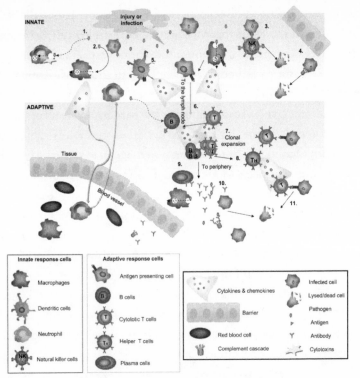

The adaptive and innate immune systems
(Figure from Garay and McAllister, 2010)

when one does, it, too, goes nuts, duplicating like crazy to produce tons and tons of copies. Most of these copies produce antibodies specific to whatever piece of invader triggered the T cell in the first place, but some stick around to become memory B cells.

With the combined attack of the antibodies, T cells, macrophages, and neutrophils, the invader is hopefully defeated, and things eventually go back to normal. But should the same invader appear again, the memory T and B cells are ready to launch a much faster response, which is how you develop immunity. Vaccines, for example, use the body's natural immune response to artificially prime your adaptive immune system: essentially, they go through the whole process of rallying T and B cells and creating that memory for a compound without you actually being infected by the pathogen. Vaccines consist either of chopped-up pathogen proteins, viruses killed through irradiation or some other means, or "live" viruses or bacteria that have been artificially handicapped, so they can't cause the same damage as their more virulent cousins. They're usually aided by an adjuvant, which is a compound that helps kick the immune system into action.

Bacteria and viruses aren't the only things that kick on the immune system; so, too, do venoms—a fact which scientists have exploited to create the most effective treatments for envenomations to date: antivenoms. Antibodies are surprisingly powerful weapons. Not only do they locate unwanted particles and signal for immune cells to seek and destroy; if the offender is an enzyme or a signaling protein, they often block its ability to do its job. It's great if someone who has been bitten by a snake or other lethal venomous animal can quickly produce enough antibodies to protect themselves against the venom toxins, but venom toxins are so damaging and quick-acting that people rarely survive long enough for the adaptive immune system to do that. By the time the body could get to the stage of activating B cells, massive tissue damage or life-threatening paralysis may already have set in. If only there was a way to have a more immediate influx of toxin-specific antibodies . . . and there is, thanks to pioneering work by scientists at the end of the nineteenth century. That's essentially what antivenoms are: antibodies that have been created ahead of time to attack specific venom toxins.

Currently, antivenoms are made by using animal immune systems as living manufacturing plants. Horses are the top choice because of their size (it means that there is a reduced likelihood that the horse will die from the dose of venom that the scientist injects, and their large blood volume allows more antibody-containing serum to be extracted at one time) and how easy they are to breed and maintain in captivity, with goats and sheep also used frequently, but scientists have made antivenoms using animals ranging from cats to sharks. To get the animal to produce antibodies, scientists inject it with a carefully determined dose of venom plus adjuvants, much like the makeup of a vaccine. If all goes well and the animal mounts an immune response without a major reaction, several more injections will follow. Exactly what is injected, where, and how often are closely guarded secrets of the scientists working at companies that produce antivenoms, and the scientists have special tricks to get the antibodies they want.

After several weeks, the animal is ready to be bled. The antibodies that will neutralize the venom are in the horse's blood, so there's really no way around this macabre part of the process. Generally, anywhere from three to six liters of blood will be drained to extract the antibodies, according to WHO guidelines. Scientists separate the animal's blood cells from the antibodies circulating in the serum by spinning the blood in a centrifuge, which causes the heavier cells to collect at the bottom of a tube while the antibody-rich plasma stays on top. Using a series of purification procedures on the plasma, they then separate as much of the "junk" (non-antibody proteins) from the lifesaving antibodies as possible. The end result is a cocktail of venom-specific antibodies that are ready to assist a human being should a snakebite or other dangerous envenomation occur. It doesn't matter that the host/participating animal is not a human: the antibodies in their blood will bind venom proteins just like our own, preventing them from having their toxic effects.

There's no doubt that antivenoms have saved millions of lives from species that were once considered absolutely fatal. But antivenoms aren't perfect: they're expensive to produce and require a steady stream of venom in order to continue injecting the animals to stimulate antibody production, as the immune response will fade after a couple of months. The antivenoms produced using animals also have limited shelf lives, so pharmaceutical companies that make them must constantly generate stocks which are then sold to hospitals, doctors, reptile breeders, zoos, and anyone else who is willing to pay the stiff cost to have antivenom on hand. While large-enough venom volumes are not necessarily hard to obtain for some snakes, it's difficult to impossible for other species, such as spiders or jellyfish, because they're either small, hard to extract venom from, or too rare. Furthermore, antivenoms usually work only for one or a few species of venomous animal, proving useless against the venom of even close relatives. In some snakes, venom composition varies so much across the animal's range that the same antivenom can't be used even on the same species a few hundred miles away!

But one of the biggest drawbacks of how we currently make antivenoms is that because they are created by using animals as the producers, they invariably introduce a slew of nonhuman proteins into the body when they are injected, any of which can set off an allergic reaction or cause other unwanted immune reactions. As many as 43 to 81 percent of snakebite victims treated with antivenom have severe side effects—but at least they don't die from the venom.

The next wave of antivenom scientists is trying to find ways to combat these problems, using new technologies to better purify animal serum, or to create treatments that work against the venoms of several species. A new field—antivenomics—has emerged, which uses cutting-edge immunological and molecular methods to clean up antivenoms. The basic idea is that antivenoms are mixtures of antibodies, many of which aren't actually targeting the most deadly toxins in the venom. Scientists estimate that less than 5 percent of antivenom proteins actually do the job intended: neutralizing venom components. By creating filtering protocols that require antivenoms to bind to toxins, scientists are able to separate out that vital 5 percent from the rest, thereby removing a large number of the molecules that might cause side effects. Similarly, they can screen current antivenoms to see how effective they'd be against other species of snake, and add mixtures of antibodies from different immunized animals together to create an antivenom that halts the venom of several species. The most ambitious of these antivenomics efforts hopes to create universal snake antivenoms that will work on all snakebites which occur in a region of interest like Africa or India.

But antivenoms are only one potential answer to the global problem of envenomation. Some scientists hope that understanding how other species resist venom's lethal power may help us devise better treatments for humans. There are plenty of anecdotal stories of venom immunity—like the now-famous honey badger, whose venom resistance was included in a dubbed-over *National Geographic* video that went viral around the world—but only a few species have been studied in any depth to determine how they survive such ordinarily fatal bites, and whether we can use their biochemical tactics to potentially save ourselves.

There are two main ways to study immunity: in vivo challenges, where a given individual is injected with a known dose of venom, and in vitro ones, where cells or bodily fluids (usually blood serum) are mixed with venom in a dish to determine if they can neutralize the venom's effects using some kind of bioassay. Some of the earliest studies simply took blood serum from a mongoose or opossum, mixed it with a normally lethal dose of venom from a cobra or other snake, then injected the mixture into a mouse to see if it would survive. These two main methods measure very different things: if an animal is resistant to direct injection, then it clearly has some degree of immunity, but all we really know is that it somehow survives. But if mixing an animal's blood with venom weakens the venom, then we know that there is something transferable that helps the animal survive, which has more pharmaceutical potential to save us down the line.

The animals on the planet that are the most resistant to venomous snakes are those that eat snakes regularly. You might be surprised at just how many species put vipers and other highly venomous snakes on the menu. Well-known examples include honey badgers, mongooses, opossums, and hedgehogs. They're not alone: there are also several badgers and weasels, skunks, and even cats who will take on venomous snakes. There are at least forty-eight species of mammals across six orders that are known to eat venomous snakes—just how many of them can resist their defensive bites remains to be seen. Most of them have never been tested for venom resistance. Those that have, though, show strikingly strong immunity to snake venoms. The *Didelphis* opossums can survive forty to eighty times the dose of viper venom that kills mice—or men. And it's not just snake predators: similar resistances have been found in species that eat other toxic prey. For example, grasshopper mice that eat bark scorpions are highly resistant to the arachnids' venoms, and are able to withstand three to twenty times the amount that regular lab mice can.

And that's the mammals. The bird genus *Circaetus* is known for dining on venomous snakes—so much so that they're commonly known as the snake eagles. Preliminary work suggests that the short-toed eagle, a member of the genus, has blood serum proteins that protect it from the vipers it feeds upon. Lizards, too, are resistant to the venomous bites of their toxic prey. There are anecdotal reports of resistance to scorpion stings in several lizard species, but the only one that has been tested and confirmed is that of the fan-fingered gecko, a native of the Middle East, to the venom from yellow scorpions. The little lizard can survive *four thousand times* the LD_{50} dose in mice—the equivalent of one hundred stings' worth of venom. The Texas horned lizard specializes in eating harvester ants, which just happen to pack the most potent venom in the Hymenoptera (an LD_{50} of 0.12 mg/kg). The lizards have developed an impressive tolerance to the toxic venom: the LD_{50} in the lizards is more than *fifteen hundred times* that of mice!

Sometimes, resistance is found in animals that are eaten by venomous animals. One of the most amusing discoveries of venom immunity in a mammal was by Texas scientists in the 1970s. They attempted to feed woodrats to western diamondback rattlesnakes, since the rats were readily available and seemed like a species the snakes would eat. Much to the scientists' surprise, the rats not only survived being tossed in as food with hungry venomous vipers, the rats sometimes even killed the snakes, biting and scratching them to death. Being good scientists, the researchers took the opportunity to study the interaction, and discovered that the woodrats could indeed tolerate the rattlers' venom, thanks at least in part to something in the woodrat's serum. They went on to purify the serum compound and study how it stops the hemorrhagic activity of the venom. Similarly, some eels that fall prey to sea snakes can handle high doses of the venom of sea kraits with few to no signs of trouble.

In addition, many species are immune to their own venoms or those of close relatives. Most snakes, for example, have a decent level of immunity to other venoms within the same family. But snakes are still susceptible to less closely related venoms—the vipers can be taken out by the elapids, the elapids by the vipers, and both fall prey to the rear-fanged colubrids (the family Colubridae), a good number of whom are snake-eating specialists. These colubrid snakes show remarkable resistance to other snake venoms; the Florida king snake, for example, has proteins in its blood that protect it from the killer activity of cottonmouth venom.

What we know from in vivo challenges is that not all snake eaters or snake prey can survive all snakebites. The opossums can resist a slew of viper venoms, from the Americas and from Africa and Asia, but they're helpless against cobras and other elapid snakes. Hedgehogs, such as the European hedgehog, are also immune to vipers. Meanwhile, the Egyptian mongoose appears to be a venom-foiling jack-of-all-trades—it can handle viper venoms and shake off elapid venoms. It's not just a little resistant, it's *really* resistant, even to the most potent venom toxins. In one study, scientists gave Egyptian mongooses increasing doses of sarafotoxin-b, the most lethal component in the venom of many African asps. The little buggers were not only able to shrug off an LD_{100} dose—the minimum dose that kills *all* of a population of lab mice—they survived when given *thirteen times that amount*. But, oddly enough, when the mongoose's blood is mixed with venoms and injected into mice, it offers no protection. They're immune to venoms, for sure, but their immunity is innate and cannot be shared.

The reason the mongoose is immune but cannot protect others with its serum is that the bulk of its antivenom adaptations are actually changes in the parts of cells that venoms target. The snakes that the mongoose is resistant to have one thing in common: they all have toxins that target nicotinic acetylcholine receptors. These cholinergic receptors (which bind the neurotransmitter acetylcholine) are a part of the neural pathways that tell cells, especially muscle cells, to contract. These snakes' α-neurotoxins—such as alpha-bungarotoxin—bind to and block the active site in the nicotinic receptors, blocking the signal to contract and causing speedy paralysis and death in unfortunate victims. But the mongoose has evolved a slightly different receptor from its mammalian relatives.

Scientists have found that five little amino acid changes in the active site are all it takes for the mongoose to be immune to the snakes' potent receptor-targeting toxins. Moreover, three of those five mutations are shared by the Chinese cobra, which, no surprise, is also immune to such toxins.

More-recent research has shown that honey badgers, hedgehogs, and pigs have independently acquired functionally equivalent amino acid replacements in the toxin-binding site of their muscular nicotinic cholinergic receptors, making them essentially immune to these elapid toxins. In all three groups, a positively charged amino acid sits where an uncharged one once did, so the toxins cannot bind. Some mongooses, too, have modified the same site, placing bulky sugars on their amino acids, which scientists think block where the toxin binds to have the same effect. Thus the resistance to snake venom α-neurotoxins has evolved independently at least four times.

It's amazing to think that these animals are able to resist such targeted and potent toxins in this way. Nicotinic receptors play vital roles in cellular communication and neuronal signaling, so it's not unreasonable to think that changes in their sequence are tightly controlled. The proteins that are targeted by venom toxins are vital to survival—they need to be able to interact with the molecules they normally do in the animal's body. So if a mutation caused enough of a change in the protein that it altered the ability of the receptor to function normally, then the animal would not be able to survive. Mutations in such vital proteins are often lethal. Not surprisingly, then, mutational immunities like the mongoose's are rare, but not unique, in nature.

The opossums, on the other hand, can confer their immunity to others because a large part of their immunity is gained from toxin-inactivating compounds circulating in their blood. Scientists have isolated serum proteins that bind venom components

such as metalloproteases (enzymes which cause deadly hemorrhaging) and prevent them from functioning. Opossums can even
pass these compounds on to their young through their milk.
Hedgehogs, too, have special components in their blood that tackle
snake venom toxins. These include macroglobulins—proteins
structurally related to antibodies (which are *immuno*globulins)—
that can completely halt the hemorrhagic activity of viper venoms,
as well as metalloprotease inhibitors like those in the opossums.
Indeed, the compounds that confer resistance in these two
distinct groups share a lot of similarities, suggesting convergent
evolution.

The weird part is that while it makes sense for snake-eating
species to have developed resistance to venoms, it makes even
more sense for species falling prey to those animals to have evolved
similar immunities. That is, after all, how evolutionary arms races
tend to work: pressure from the predator leads to adaptation in
the prey to escape, which leads to adaptation by the predator to
succeed in its catch, and so on and so forth. But far fewer prey
species have developed venom resistance than expected (those
Texas woodrats notwithstanding), and they tend to be less resistant than the predators of venomous species. And while there's
plenty of evidence for rapid evolution of venom toxins, there's no
evidence for rapid evolution of antivenom proteins in response
in prey species. Instead, that rapid evolution has been detected in
antivenom serum proteins from the predators of venomous taxa,
which suggests that the venomous snakes may have developed
their potent venoms at least partially *for defense*.

The lack of resistance in prey species also suggests that there
must be evolutionary constraints which make it difficult to develop
resistance, or it would be more commonly distributed. Perhaps
antivenom compounds are just too expensive to be practical in
most prey species, given the odds of death by venom as compared

to other ends. There's a lot to be learned about how coevolution works. But through studying venom-resistant species, scientists can peek behind the curtain and learn even more about the rules that govern natural selection.

Though we don't understand exactly how different species have evolved such stunning resistances to venoms, scientists may still be able to use their physiological insights to develop similar proteins and provide a cheaper alternative to antivenoms altogether. It's not so hard to imagine a universal snakebite treatment that uses modified serum proteins from species like opossums to combat a wide variety of deadly venoms.

But even if such treatments work, they would still entail animal proteins being injected into human blood, which means it's likely that there will be adverse reactions. Some of the most interesting endeavors to confront the global challenge of snakebite are those that take an entirely different approach to the problem: if a large part of the issue is that the treatments we use to combat the snake toxins don't come from humans, couldn't we just figure out a way to have a human produce antibodies instead? Steve Ludwin is trying to do just that.

Steve Ludwin isn't a scientist. He's not a doctor or a researcher. In the late 1980s, he dropped out of college in New England and moved to London to seek a career in music. He's written songs with Slash, been a member of several bands, and even went on a date with Courtney Love once. But to most people, he's not known for his rock 'n' roll career. He's known as the guy who injects himself with snake venom.

"It was about 1988, 1989 when I first started experimenting with venom," Steve explained to me. It was a time before the Internet, before there were books or papers on how to self-immunize or Facebook groups dedicated to the practice, as it is referred to within the community, of "SI'ing." Modern SIers have a myriad of resources and a community with whom to discuss various aspects, from dose to species. Steve had none of that. "It was just all very instinctual what I was doing. I kind of just made it up as I was going along."

Steve was always an avid reptile lover (or "herper," as it's known in the world of reptile enthusiasts, as the study of reptiles and amphibians is called herpetology after the Greek *herpetó*, or "creeping thing"), and fell in love with snakes at a young age. When he was nine years old, he met Bill Haast, the director of the Miami Serpentarium, who is considered one of the pioneers of SI. Haast, whose facility took care of hundreds of snakes, was one of the most well-known venom scientists of the past century. The Serpentarium was one of the largest producers of venom for pharmaceutical and antivenom research for decades, but Haast also had a side project: self-immunization. He did it as a means of self-preservation, but also because he was simply curious whether the usual process of antivenom production could be repeated in a human. He began with cobra venom in 1948, but as time wore on, he added more and more species, eventually injecting himself with mixtures of a few dozen venoms at once. Even when Haast closed the Serpentarium in 1984, after a freak accident in the alligator pit led to the death of a young boy, he continued experimenting on himself with the snakes he still kept and milked for research. He did appear to have incredible immunity: he was bitten more than *170 times* in his career, and though he had some close calls, he would always recover. His intentional venom injections saved him over and over, or so he believed, and he was so confident about his antibodies that when he was informed of life-threatening bites in his local area, he donated his blood to victims if antivenom wasn't available (many say he saved several lives by doing so). In media interviews, Haast explained that in addition to his immunity to venoms, he was in remarkable health, and he believed that the venom was actually improving his immune system. When he was eighty-eight years old, Haast said that if he lived to the age of one hundred, it would be evidence that the venom was a health booster. And he did.

Back in the 1970s, when Steve visited the Serpentarium, Haast made quite an impression on him. "I remember thinking, 'Wow, you can inject yourself with snake venom and become immune to it? That's pretty cool.'" By age seventeen, Steve had decided to follow in Haast's footsteps. He describes the decision as a "moment of clarity," that he simply *knew* that he was going to inject snake venom into his body. He thought about which snakes, how much, and how often. Soon enough, he got started, and he has been injecting himself ever since.

Every couple of weeks, Steve injects a cocktail of six to eight different snake venoms into his veins. He's tried dozens of species, from hemotoxic vipers (which feel "like you've put Tabasco sauce under your skin") to neurotoxic elapids, including cobras ("you don't get the pain element from snakes like that"). Sometimes, Steve said, he even feels energized by his injections. "It's not a high, but it's like this rush of like, *Wow!*" he told me. "I feel like I'm twenty-four years old again."

Most people who SI claim it's for the immunity itself—they are avid keepers of venomous animals, and want to ensure that they have an extra layer of protection should a bite occur. But I think there's more to it than that. All SIers whom I have encountered have immense pride in their practice. They believe they are on the cutting edge of science, even though they might not personally be contributing to new technologies. The way they see it, they know something that scientists don't, and their very survival in the face of venomous bites proves it. They boastfully handle their toxic pets without hooks or other protective measures. Their Facebook profiles are littered with photos of them kissing cobras on the head or with vipers wrapped around their necks, smugly defying conventional recommendations about safe reptile keeping.

Many in the venom community, from scientists and medical professionals to reptile keepers and enthusiasts, have spoken out loudly against SI'ing. Some of the most well-known scientists condemn the practice wholesale. But Steve doesn't understand that. "I don't see why it's treated as such voodoo science. It's facts. A horse can become immune to snake venom and develop antibodies, and so can a human." Though, if he's truly honest, that's not really why Steve does it. "I'm not doing it to protect myself against snake venom, I'm doing it because I'm intrigued by it," he said. "I've always had this feeling that something positive would come out of it . . . While everyone is out exercising their arms, making their muscles bigger, I kind of see it like I'm exercising my immune system."

A growing number of people, largely teenage boys and young men who already keep snakes and other reptiles, have started SI'ing with their venomous pets (or "hots," as they call them). SIers insist that anyone who keeps or works with hots should immunize themselves against the species they interact with, and they feel that they're on the forefront of an important movement. But most keepers think that SIers give reptile owners a bad name, likening SI to a macho cult. The SIers, in turn, simply video their bites (by notorious species like black mambas) for all to see and erect a particular finger to anyone who calls them crazy.

You might think that after more than twenty years, Steve would welcome the like-minded company of these young, rebellious SIers, but he is quick to discourage anyone from joining the club. "It's obviously dangerous," he said. When he conducted an online discussion forum to talk about the practice, he added this disclaimer:

> I do not advise, recommend or condone any person to self-immunise or immunise any other person or persons with snake venom. Immunisation with snake venom of any type, whether carried out by injection, ingestion or any other means, is an extremely dangerous and experimental activity which is likely to result in serious injury or even death and should not be carried out by any person whatever the circumstances.

While he wishes that some of the scientists and doctors would take SI more seriously from a research perspective, he has no patience with the SIers who appear to be doing it for the sake of bravado, referring to some he's seen as "braggy." He said he doesn't participate on the SI Facebook page or reply to the constant stream of e-mails from those who want his tips because "it's an accident waiting to happen . . . You tell them what to do, they fuck up, and then you're responsible."

And Steve would know all about that; for him, SI'ing hasn't been a smooth ride. "I have that stupid rock 'n' roll mentality of 'Whatever, just do it.' So I've had so many accidents," he said. Early on, his arm would swell like a balloon. Part of that was because he was injecting hemotoxic venoms ("I literally was so stupid that I didn't really know the differences between the venoms back then"), but he also readily admitted he often used too much. "I always kind of thought just get more in, more in, more in," he explained, but now he realizes that even small amounts still stimulate the immune response.

Steve has had necrotic holes in his flesh from his self-immunizations. When one particular injection went wrong, Steve ended up in the hospital with nurses and doctors telling him he was going to lose an arm or die—but he didn't. He's only actually been bitten once, by an eyelash viper—an accidental nip from a pet that moved a little faster than he anticipated. Since he'd been self-immunizing, he decided to wait it out to see if he was truly immune. He luckily survived, but it was the worst pain he'd ever felt. "It felt like someone had basically slammed a sledgehammer onto my finger," he said, "and that's what it stayed like for eight hours." But in the end, he doesn't regret the past twenty-six years he has spent self-immunizing. Though it isn't perfect, he said, it does work to generate immunity: "I've put lethal amounts into a syringe with two people watching and I've injected it to prove the point that my immunity is pretty good to this stuff."

And, like Bill Haast, Steve believes the benefits from his injections go beyond protection from snakebite. He noted that snake venoms have been a part of traditional medicine in a number of cultures throughout history, and though such folk remedies aren't always spot-on, they tend to be based in real biological activities. There's also his overall health: "I don't get colds. I don't get sick. I don't get the flu," he said confidently. Not that he never gets sick: "I had food poisoning a couple weeks ago, which sucks," he added. "Snake venom doesn't seem to do anything for food poisoning, I can tell you that."

The similarities between Steve's experiences and those of other SIers are hard to ignore. "Bill Haast always said the same thing, in the few interviews he did. He said he's the picture of health, never been sick a day in his life," Steve noted. "And you start putting two and two together, and you start thinking 'There's something in this, and it needs to be studied.' But it isn't being studied."

Well, self-immunization *wasn't* being studied—until a couple of years ago, when a video of Steve's caught the attention of researchers at the University of Copenhagen. Steve was overjoyed as he told me how scientists are now studying his blood, hoping to use his antibodies as blueprints to make clean, human-derived antivenoms. The five-to-seven-year project could be the next big breakthrough in treatment technology. Steve isn't being paid a penny for his contributions; he just wants credit when his antibodies start saving lives. He still hopes that future research will investigate the potentially immune-boosting aspects of snake venom, but for now, the research into his blood has brought a new energy to Steve's experimentation. "It all feels really good now, like there's a purpose. Because to be honest, when I was doing it when I was younger, I had no idea what exactly I was doing it for. I didn't understand it back then. Now it has validity."

Since our primate ancestors often fell into the category of prey rather than predator of venomous species, it isn't surprising that we are not immune to venomous animals—not even somewhat immune in the way Steve is. But there is a small amount of evidence that we can gain some level of resistance thanks to our adaptive immune system. Unfortunately, that's the same system that makes normally harmless venoms, like the toxins in honeybees, potentially lethal due to allergic reactions.

No one really understands why we have allergies; it's an immunological mystery that scientists have been debating for centuries. You can think of allergy attacks as moments when your immune system overreacts. Allergies are, by definition, "hypersensitive immune responses." The thing you're hypersensitive to is the allergen, and it can be anything that the body can recognize using the antibody-making immune pathways. Allergic reactions don't happen the first time you are introduced to an allergen; instead, your body takes an immunological picture so it can remember the allergen later. When the allergen reappears, your immune system goes haywire, sending out massive armies of antibodies—as our immune systems should. But for some reason, certain antigens cause the body to send out IgE antibodies instead of the more common IgG ones. IgE antibodies are problematic in themselves. They make up only 0.001 percent or less of our antibodies, and for good reason: they cause massive releases of histamine and other inflammatory compounds, which can lead to whole-body reactions called anaphylaxis. Anaphylaxis can be as benign as a dip in blood pressure, or as deadly as cardiac arrest. Given that IgE antibodies are so potentially troublesome, scientists have sought to understand their role in the immune system. And that's the weird part: they don't seem to have much of a benefit—only the cost of allergic reactions, which some 20 to 30 percent of people experience.

There isn't a whole lot of evidence to explain why humans have IgE antibodies in the first place. It's a puzzle to immunologists. Why make a class of antibodies that do more harm than good? IgE antibodies have to have performed some sort of useful function at some point in our evolutionary history, or the constant cost of allergies would have gotten rid of them. Some have suggested that they may play a role in the fight against parasites, and that now, in our more sterile lives surrounded by Purell and penicillin, we have removed the IgEs' foes, so we see only when they malfunction. There is some evidence to support this hypothesis, but it still suggests that allergies are a side effect and not the goal of IgE production. And it can't explain why certain things tend to be more allergenic than others—are our parasitic defenses so poorly tuned that they mistake pollens, foods, drugs, venoms, and metals for parasites? Other scientists suggest these vexing antibodies might have a more intriguing use: to protect against toxins, including venoms.

The Toxin Hypothesis was first presented by the elusive scientist Margie Profet in 1991. Though her degrees are in physics, math, and philosophy, Profet shocked the immunological world with her radical idea that allergies could have evolved in their own right, and not as a side effect of other processes. "The evolutionary persistence of the allergic capability, despite its physiological costs, implies the existence of an adaptive benefit for this capability that outweighs the costs," Profet explained. "This undermines the view that allergy is an immunological error . . .

"The specialized mechanisms that collectively constitute the allergic response appear to manifest adaptive design in the precision, economy, efficiency, and complexity with which they achieve the goal of producing allergy," she wrote.

The Toxin Hypothesis consists of four main arguments: First, that toxins are ubiquitous and cause acute damage, which essentially speaks to evolutionary motive. If toxins are common and deeply harmful, then it only makes sense for our bodies to have developed defenses against them. Furthermore, Profet noted that most toxins are both acutely damaging *and* cause long-term damage. Many, for example, are mutagens that can induce cancers.

Second, the types of physiological activities that toxins perform are known to trigger allergic reactions. For example, many toxins covalently bind to serum proteins, an activity that also often triggers allergies.

Third, most allergens are either themselves toxic substances or carrier proteins that bind to smaller toxic substances. Venoms, for example, are in and of themselves acutely toxic, but even some of the more harmless-seeming allergens can carry toxins. Hay, for example, can be laden with fungi-derived aflatoxins that can cause acute liver failure.

Finally, the Toxin Hypothesis states that allergic symptoms could be construed as helpful in the case of envenomation or poisoning. If the body has refined its reaction to IgE, then allergic symptoms must be beneficial. Indeed, behaviors like vomiting, sneezing, and coughing could expel toxins, and drops in blood pressure could slow the speed at which a toxin moves through the body. Even the release of heparin, an anticoagulant, during allergic responses could be construed as a way to fight against the coagulatory actions of many venoms.

According to Profet, allergies are essentially the adaptive immune system's last and most desperate line of defense against toxins, including venoms. That allergies get worse with each subsequent exposure to an allergen isn't an immunological mistake: it's a key part of the design. Since the risk from multiple exposures to a toxin can be cumulative, the more you're exposed to a toxin, the more important it is to get rid of it faster the next time. That's not to say that allergies, in their current form, aren't a bother. Billions of dollars are spent every year treating the runny noses, watery eyes, and itchy hives that are brought on by a diverse set of triggering substances. But to focus on such nuisances is to ignore the bigger picture, supporters of the hypothesis would argue. Allergies are seen as a nuisance, they'd point out, only because we don't appreciate how often they have saved our butts.

The Toxin Hypothesis won Profet a MacArthur Foundation "genius grant" in 1993, but even now, the scientific community has not entirely warmed up to it. Scientists continue to say that there's simply no *experimental* evidence to support it. Some suggested (as Profet did) that lower cancer rates in allergy sufferers might be because allergic reactions expel carcinogens, but that alone wasn't enough of a smoking gun. After all, an overreactive immune system could attack all things more vigilantly, making it hypervigilant against cancer, too. If the Toxin Hypothesis proves correct, then the allergic response itself specifically has to be somehow beneficial.

It would take twenty years before Profet's radical proposal was supported by experimental evidence. In 2013, scientists showed that small doses of bee venom and the subsequent allergic pathways triggered serve to protect mice against fatal doses of venom later on. The most convincing evidence was that mice genetically modified to lack one of the steps in the allergic response (either IgE antibodies, IgE receptors, or the mast cells that express the receptors) did not benefit from pre-exposure, directly linking the IgE response to the protective effect. Then scientists tried a much more potent venom, that from the Russell's viper. As with the bee venom, the primed IgE response protected the mice.

There's still a lot to be explained even if the Toxin Hypothesis holds up to further scrutiny, including how a fine-tuned immunological response system fails in the case of anaphylaxis. But it's certainly a compelling hypothesis to explain why our bodies react the way they do to toxins, especially venoms. And it lines up with what we know of venomous creatures; specifically, that their potent toxicity has a way of influencing the species around them. While we may not have developed venom-proof proteins or antivenom compounds in our blood, our ancient, smaller ancestors—and other preyed-upon species, like mice—may instead have developed a complex immune pathway whose entire purpose is to deal with life-threatening toxins like those in venoms. And if the Toxin Hypothesis is true, then scientists may not need to look so hard to find potentially lifesaving therapeutics. The secret to surviving envenomations may be hiding in plain sight, disguised as an allergy.

There's no doubt that better treatments for life-threatening envenomations are desperately needed. It's estimated that more than 400,000 venomous snakebites occur annually, killing as many as 100,000 people. And then there are the deaths from other venomous animals—spiders, scorpions, jellyfish, and the slew of killers I explored in the last chapter. But the future of antivenom science is bright; we have many promising roads ahead, whether we turn to Profet's insights, immune animals, SIers, or antivenomics. In addition, the more we learn about how venoms work on a molecular level, the better equipped we are to find new ways to combat these toxins—even the ones that aren't lethal. After all, many of the venoms that don't kill still manage to produce unimaginable pain.

4

TO THE PAIN

"Life is pain, Highness. Anyone who says differently is selling something."

—THE MAN IN BLACK (*The Princess Bride*)

The entomologist Justin Schmidt wrote that the sting of the bullet ant is "Pure, intense, brilliant pain. Like walking over flaming charcoal with a three-inch nail in your heel." According to Schmidt, it is the most painful insect sting in the world. And he would know—he's been envenomated by seventy-eight species and forty-one genera of the order Hymenoptera (bees, ants, and wasps) in the process of developing the Schmidt Pain Index, a colorful and cheeky scale which describes the pain of stings from 0.0 (harmless) to 4.0 (unfathomable agony). On that scale, the bullet ant scores a 4.0+—the only species to score greater than a 4.0. The much milder sting of the bald-faced hornet scores a 2.0, as it is, to Schmidt, "Rich, hearty, slightly crunchy. Similar to getting your hand mashed in a revolving door." Meanwhile, the tarantula hawk, the largest wasp in the world, packs a 4.0 sting, which in Schmidt's words is "Blinding, fierce, shockingly electric. A running hair dryer has been dropped into your bubble bath."

The bullet ant's highest score comes as no surprise given that its common name refers to its sting as being the pain equivalent of a bullet wound. According to people who have been stung, not only is the agony intense for three to five hours, it takes a full day to subside completely, with common "side effects" such as trembling, nausea, and sweating. So naturally, bullet ants were one of the animals I was most excited to see on my journey into the Peruvian Amazon—from a safe distance, of course.

Bullet ants are the bane of many an Amazon tourist, but to the Sateré-Mawé people of Brazil, they are a part of their cultural heritage, playing a pivotal role in the initiation rites for young Sateré-Mawé men to become warriors. To prepare for the initiation ceremony, village elders carefully collect about one hundred bullet ants from the forest and drug them with an herbal sedative. They then weave the ants into gloves made of leaves, with their stingers facing inward. When the ants awaken, they are enraged, ready to sting anything that comes into contact with them. Before a boy can call himself a man, he must wear the bullet ant gloves, subjecting himself to hundreds if not thousands of stings, which make his hands swell into clubs while his body shakes in pain.

Even though the ritual is still practiced today, accounts in English vary as to the details of the rules. Some say the gloves have to stay on for ten minutes—others, thirty. Though you'd think once was enough, tribesmen undergo this ceremony up to twenty-five times throughout their lives, starting at age twelve. Their reason for the repetition? Some who have witnessed these ceremonies claim that the boy cannot cry out or shed a tear as the venom takes effect, and if he cries, he must repeat the ritual again. Others have said that rather than being forced, the young men voluntarily repeat the painful ritual to gain respect and leadership.

Bold actors and filmmakers, among many others, have attempted to survive the same ceremony. In one, the Aussie comedian Hamish Blake lasts only a few seconds in the gloves, and ends up in the hospital hours later after collapsing from the unrelenting agony. Pat Spain, a *National Geographic* presenter, made it an entire five minutes, only to descend into incoherence, unable to speak or stop shaking for hours after the stings. He was still incapacitated five hours later when his arm was immersed in ice water to relieve even the slightest bit of the pain.

The TV personality and adventurer Steve Backshall wrote about his attempt to withstand the sting of the bullet ant for *The Sunday Times* in 2008. It wasn't the ten minutes in the gloves that was the problem, he said ("it wasn't that bad: pretty unpleasant, but bearable"), but the hours afterward:

> First, I started wailing, then, once that had passed, the floodgates opened—deep, guttural sobbing, uncontrollable shaking, writhing, convulsing. You could see the neurotoxin kicking in, my muscles starting to palpitate, my eyelids becoming heavy and drooping, my lips going numb. I started to drool, and suddenly I wasn't responding to anything at all. My legs wouldn't hold me up, and our doctor was shouting at me to keep moving and not to give in to the urge to lie down and let it take me.
>
> If there'd been a machete to hand, I'd have chopped off my arms to escape the pain.

According to Backshall, it was a full three hours before the pain began to "ease a little."

Bullet ant stings are so insanely painful because unlike snakes or spiders, which use their toxins to capture or digest prey, the little ant has one goal: defense. The crippling agony that its victims experience is thanks in large part to one small peptide called poneratoxin. As much as one microgram (1µg—though a tiny weight, it's the equivalent of us carrying around a pound or so) of poneratoxin can be stored in the ant's venom reservoir, ready to be injected into its next hapless victim. The compound alters voltage-gated sodium channels in neurons, causing the nerve cells to go haywire. Muscles lose control, and the neurons that conduct messages of pain are relentlessly stimulated. This excruciating pain lasts for hours, as the cells in our bodies are helpless against the small peptide toxin. It's the perfect message: Back off! The pain is enough to convince any potential threat that messing with *these* ants is a grave error. Just thinking about the physiological effects makes me shudder.

Of course, of all the animals that could have snuck back to Hawaii with me from the Amazon, it was none other than a bullet ant that I found when I washed my mud-and-sweat-soaked laundry. I stared at it in disbelief. A bullet ant. *In Hawaii.* Thankfully, I found the little devil at the base of my washing machine when I took out a load of laundry—dead, presumably. The ant had been somewhere in my clothes. The clothes that I packed *bare-handed* into that suitcase. The clothes I removed from said suitcase, *barehanded*, to put into the wash. *How many chances did this ant have to sting me?* I poked the animal with long metal forceps I just so happened to have nearby (I use them to feed clams and crabs to my pufferfish—besides, you never know when you might need them!), just to be sure it was really dead. Nothing. *Good.* I let out a deep breath and carefully pulled it out of the machine.

It was small for a bullet ant, less than three-quarters of an inch long. It looked so . . . harmless.

Just a week before, I had held a similar ant in similar metal forceps while Aaron Pomerantz, Frank Pichardo, and Jeff Cramer set up their camera equipment in Aaron's small dorm room at the Tambopata Research Center. Jeff is a world-renowned photographer who works for the company that owns the center's lodges, while Aaron is his biologist-for-hire, aided by the local photographer and guide Frank. That ant was alive and extremely pissed off, exactly the way the boys wanted it so they could get a high-resolution macro shot of its terrible stinger. They fussed with flashes and lenses while I, with extreme trepidation, held a writhing, furious ant. The ant that had traveled home with me had no such vigor after the spin cycle. Ironically, we had joked about letting one sting us just to know what it feels like ("For the book," Aaron said, "it would make a great anecdote!"), but all of us wussed out. I imagine I would never have lived it down if I spent two weeks dodging bullet ants in the Amazon only to feel the "pure, intense, brilliant pain" a whole three days after returning to Honolulu.

I placed the ant on a piece of paper, and left it on my coffee table as I headed to bed. When I went to place my unexpected keepsake in a small ethanol jar in the morning, the bullet ant was gone. Perhaps blown by the wind? I couldn't find it on the floor. Eaten by something in the night? I can only assume so—I mean, there's no way the ant was alive after all that it went through, right?

Right?

To this day, I still get nervous every time I sit on my couch.

You've most likely been envenomed at least once in your life, and, luckily for you, by a defensive venom user: the bee. You also likely learned the lesson that venom is meant to teach you: Stay away. Pain is one hell of a way to turn off a predator, and venomous animals have found all kinds of ways to induce it. Many use our own nerves against us, turning on our body's neuronal signals that warn our brains of damage or heat without actually causing

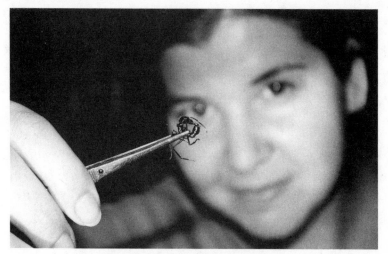

Examining a bullet ant at the Tambopata Research Center in the Peruvian Amazon (Photograph by Aaron Pomerantz)

A bullet ant's impressive stinger (Photograph by Aaron Pomerantz)

much in the way of injury. Bee venom, for example, is largely composed of a compound called melittin. Melittin takes the molecules that make up our membranes and selectively chops them into signaling compounds which turn on heat-sensing neurons in our peripheral nerves. So when you feel like a bee sting burns, you're entirely correct—the melittin causes our nerves to believe they're on fire. We know that wasps and jellyfish, too, use the same biological pathway to turn up the pain, though they do so with different chemicals. But personally, I am more interested in a different group of defensively venomous animals: I spent five and a half years studying the venomous lionfishes.

If you follow the reef flat at the mouth of Kane'ohe Bay out toward the sea, you'll suddenly find yourself in deep water. Without warning, the ground below you gives way to a near-vertical wall that plunges more than one hundred feet. Slowly descending this wall, I found myself in awe of the intense beauty facing me: coral heads surrounded by colorful fish; the tip of a moray eel's gulping jaws poking out from a small hole; a bright blue sea slug only as big as my pinky nail. As I neared the base of the cliff in my scuba gear, I headed to the south, searching for a small, pass-through cave that serves as a perfect habitat for lionfish. Like most lionfish, the Hawaiian species tend to hide out under ledges or in crevices during the day, preferring to actively hunt at night. I was hoping that if I looked hard enough, I could find one of these elusive beauties. I needed specimens to bring back to the lab in order to study their venom, to learn more about how the toxins in these stunning fish evolved.

Some people are very cavalier about diving, but I took my former instructor's words of wisdom to heart: "Every time you dive, you are experimenting with your body." One hundred feet down, I was under four times the pressure I would be at the surface. The number one rule of diving is to keep breathing, because if you hold your breath at depth, the air will expand as you rise and rupture your lung tissue. As I entered the cave and darkness surrounded me, I reminded myself to just keep breathing. *In. Out. In. Out.* My heart started to beat a little faster. The walls seemed so close, like I was only inches away from being pinned, stuck forever in a watery grave. Struggling against claustrophobia, I reminded myself that I was perfectly safe, that the cave was more than big enough for me, that I had plenty of air. I breathed in. Focusing my flashlight on the surface above me, I began to look for my fish. I checked my depth: 109 feet. As swells rocked me back and forth, I went to place my hand on the side wall to steady myself as I hunted.

What I didn't see was that a rock on the wall wasn't a rock at all. It was a devil scorpionfish, *Scorpaenopsis diabolus*, ten inches of cryptic body armed with a row of venomous spines, hidden in plain sight. The word *devil* says it all. My breath caught in my throat.

Scorpaenopsis diabolus are brutish fish. Like most of the other Scorpaeniformes, they are ambush predators, practicing a hunting strategy aided by their rock-like appearance. Rugged and hardy, these fish rarely retreat in the face of danger, assured of their spiny defense. Many fish have spines, but the Scorpaeniformes—a large and diverse fish order with more than sixteen hundred species, including the scorpionfishes, lionfishes, and the notorious stonefishes—have perfected their use. These spines don't just spear potential predators; they are lined with venomous tissue. This tissue is tucked into grooves along each of the fish's dozen dorsal spines, covered only by a thin layer of skin. In the face of danger, devil scorpionfish stiffen their bodies, erect their spines, and wait for the potential predator to realize they are messing with the wrong fish. But, like many toxic animals, they also have signaling coloration. Give a devil scorpionfish time to react, or annoy it enough from a distance, and it will flip its pectoral fins upside down to reveal the bright red, orange, and yellow markings that lie underneath. This is your warning: Proceed with caution, danger ahead. Ignore this display and you'll discover just how generous a signal it was.

As the spines penetrate flesh, the thin tissue covering the venom gland grooves is pushed back or torn to shreds. A complex chemical cocktail of proteins and peptides leaches from the venomous tissue and enters the bloodstream, spreading inward from the fingers. Some of the components act on the circulatory system, ensuring that the venom is pumped farther into the body. Other portions target nerve cells. Acting at the communication junctions of nerves, these venom components cause sudden influxes of calcium and sodium across cell membranes, leading to a massive release of the neurotransmitter acetylcholine. Acetylcholine was the first neurotransmitter discovered, earning the German biologist Otto Loewi a Nobel Prize. (He demonstrated that electrically stimulating nerves to one still-beating frog heart released a chemical that alone was sufficient to change the rate of a second heart's beat.) As one of the primary communication molecules between cells, acetylcholine has many functions, including the stimulation of muscles and sensory neurons. Those sensory neurons send signals to the brain that something is very, very wrong, which our brains decode and feel as pain. Fish venoms trick these cells into firing without provocation, when there isn't cold, heat, or an injury to trigger them.

As the venom spreads through the body, the first thing we feel is intense, unimaginable pain—without any visible cause. It fools our nervous system into feeling as if our tissues are dying, long before any real damage occurs. Using our nerves against us, the venom triggers radiating agony. The systemic response is the real killer, though: it can be so intense that it sends the body into shock. Blood pressure and heart rate plummet, disabling or knocking out the victim—a serious issue confronting me, considering that in that cave, I was in an environment far from medical assistance, where, without my gear, I couldn't breathe.

In a 1959 article for the journal *Copeia*, the esteemed biologist Heinz Steinitz described an unfortunate encounter with a lionfish, the prettiest of the Scorpaeniformes. Flamboyant red and white stripes have made them incredibly popular aquarium fish, and a favorite of snorkelers and divers who spot them in the wild. Steinitz had been swimming off a beach in the Red Sea when he came upon a young lionfish resting on the sandy bottom. Most lionfish hunt along reefs or hide in rocky crevices during the day, so he was intrigued by the fish's odd behavior. As he got closer and stretched out an arm, he noticed the fish rolled to point its dorsal fins at his hand. The scientist in him just couldn't resist. He tried approaching several times, marveling at how the fish turned this way and that to always keep its dorsal spines in prime position. Then everything went wrong. "It moved more rapidly than I was able to, and, worse, more rapidly than I had calculated. My experiments came to a sudden end."

Within ten minutes, pain began radiating from his stung finger. "I was tortured by pains beyond measure, and yet the pain was still growing more intense . . . I tried to sit down, to lie on the ground, to stand still. The pain would not let me. I had to move on, to run about. It is a strange experience recognizing quite lucidly that nothing fatal has happened to oneself, and feeling at the same time, that this was much worse than anything previous. In fact, it is just short of driving oneself completely mad." Steinitz was lucky. He was in shallow water, close to shore, able to bring himself out of harm's way very quickly. And he was stung only by two of the animal's spines, and lionfish are considered one of the least venomous Scorpaeniformes.

Although venomous fish don't generally sting to kill, death can be an unfortunate side effect. Those that have perished from the sting of venomous fish tend to do so after hours of torture, similar to the hours of maddening pain from a bullet ant sting. Though fatalities are rare, the gruesome track record of stonefish, for example, has earned them the title of the most venomous fish in the world. Their name reflects their incredibly camouflaged appearance (according to some, they are not just cryptic—they're "repulsively ugly"). The medical literature contains terrible accounts of frighteningly painful stings from stonefish. "Within ten to fifteen minutes the victim either collapses or becomes delirious and maniacal, raving and thrashing about," notes one such paper. "If stabbed while wading, it generally takes three or four men to hold him and get him to shore without drowning." Another account describes the pain as so "fearful" that "a person may become almost demented, and . . . may die."

I was lucky down in that cave. The scorpionfish saw my hand coming, and with a flash of the colorful undersides of its pectoral fins, warned me of the mistake I was about to make. It gave me just enough time to slow my hand, and instead of stiffening and bracing itself as scorpionfish usually do when threatened, it swam out from under my grasp.

Scorpaeniformes aren't the only fish to boast a venomous defense. Stingrays, too, are armed with painful venom. With any of the venomous fish, a poorly placed hand or foot is a grave mistake. The reward for such a blunder is what many describe as unimaginable pain. Accompanying it can be sweating, nausea, vomiting, changes in heart rate, and shock. Though they're excruciating and horrible, most fish stings don't result in death, because death is not what the animal intends. In the first written account of death by venomous fish, the fatal blow was actually inflicted by a human, but it involved a stingray's venom, and it is truly legendary. As one version of the story goes, the seer Tiresias told the mythical warrior and traveler Odysseus that his death would come from the sea. Yet Odysseus had been told by an oracle that his son would kill him, so death by sea seemed unlikely. When he returned home from the Trojan War, Odysseus took precautions against his son Telemachus, completely unaware that during his long journey home, he'd fathered another son by the witch Circe. That boy, Telegonus, was desperate to meet his father and found his way to Ithaca, where Odysseus had just returned. Hungry, Telegonus tried to poach from a herd of livestock, and the owner—none other than Odysseus—came to their defense. Unaware of whom he was fighting, Telegonus stabbed his father with his unique spear: a spear tipped with the venomous barb of a stingray. As the story goes, Odysseus died slowly from the wound, lamenting his excruciatingly painful demise, realizing only at the bitter end that both prophesies had come to pass.

While Odysseus' death became legend, the person most of us think of when it comes to death by venomous fish is Steve Irwin. It was morbidly fitting that he was filming a TV movie called *Ocean's Deadliest* when the iconic Australian TV host died at the age of forty-four. Though he was known for wrestling alligators and tackling the biggest, baddest species on the planet, he succumbed to an animal no one except Tiresias could have predicted. Like the warrior Odysseus, Irwin was killed by a stingray barb. We think of stingrays as some of the most docile fish on the planet. We feel so safe with them that tourist hot spots sell stingray experiences in which people stand in water as these dangerous fish swim up and over them, feeding from their hands. Aquariums around the world put them in touch tanks, allowing visitors to feel their smooth, soft wings. Irwin probably didn't even think twice about diving in that day. But in an instant the gentle giant turned, flicked its tail, and pierced his chest. Steve Irwin shouldn't have died. If the barb from that stingray had embedded itself deep in his muscle, he might have survived, although he still would have suffered tremendously. Instead, the barb struck through the small gap between his ribs and mortally ruptured his heart.

Despite their divergent evolutionary relationship (cartilaginous fish like the ray parted ways with bony fishes some 420 million years ago), stingrays are armed with a venom very similar to that of Scorpaeniformes. Both cause excruciating pain through nerve cell activation, using proteins as key toxin components. Instead of a row of dorsal spines, though, stingrays are armed with a single barb attached to their tail. The barb is a formidable blade, and its serrated edges tear through flesh like a steak knife. In case this sharp, several-inch-long serrated barb isn't enough of a deterrent, the stingray's dagger is dripping with potent venom. The hard barb is surrounded by venomous tissue that ruptures and leaches toxins when the barb enters flesh. Because the barb is jagged, it's almost impossible to remove without causing even more damage unless you're a trained surgeon, which leaves sting victims the terrible choice of leaving it in and allowing more and more pain-inducing venom to enter their system, or tearing apart their own flesh and risking excessive bleeding to rip it out.

To make matters worse, all venomous fish make getting stung all too easy. While the scorpionfishes and stonefishes look like the reefs and rocky shores they inhabit, the flattened bodies of stingrays and their habit of burying themselves in the sand can make them very hard to spot, especially for beachgoers wading into shallow water for a quick dip in the sea.

———

Venomous fishes not only shed light on the evolution of painful toxins, they also allow scientists to study the selection pressures that maintain toxicity. One might expect to see the kind of coevolution between these fishes and their predators that we see with the cobra and the mongoose—but so far, no one has looked. No one even knows which predators to look at, as scientists are unsure what species feed on venomous fishes—if any do. I decided to work on venomous fishes because so little is known about venom evolution in these fascinating species. What intrigued me most about the order Scorpaeniformes is that although the group includes the most venomous fishes in the world, there are also a lot of less notorious members. What's most perplexing for scientists like me is that the various venomous groups—scorpionfishes, lionfishes, and stonefishes— aren't one another's closest relatives. Even though they all possess unbelievably similar protein toxins that are derived from the same genes, stonefishes are rather distantly related to the other venomous fishes, and in between are numerous lineages of harmless ones. Did venom develop more than once in this order? It seems doubtful, given that the toxins found in this group are unlike any others known to man, and don't even resemble any known proteins of any kind. But then, in the harmless species, where did the venoms go?

I discovered that some of the nonvenomous Scorpaeniformes still have the genes for toxins; they just produce less of them in their tissues, and over time, random mutations have decreased their potency. It's the same pattern that the Australian venom scientist Bryan Fry found when he compared venomous snakes and lizards to their innocent relatives. His discovery that even nonvenomous snakes produce small amounts of venom proteins revolutionized our understanding of venomous reptiles. Instead of two independent origins of venom—one in lizards, one in snakes—Bryan argues that all venomous reptiles and their nonvenomous kin can trace their lineage back to a single venomous ancestor. "Nothing in evolution is ever really lost," he explained.

Scientists now think that the ancestor of the reptilian lineage that contains the venomous reptiles, the Toxicofera, developed venom-like proteins in its saliva, perhaps to protect against bacteria and other microscopic invaders. Small tweaks over time led to a novel use for these toxins: feeding. As the toxicoferans diversified, some found other ways of capturing prey (constricting snakes, for example) or switched to vegetarian diets (as the iguanas did), and the pressure to maintain toxicity slackened. The end result is a branch of reptiles with only a few strongly venomous groups interspersed among much less lethal branches.

If the same is true for the Scorpaeniformes, it may explain why they are one of the most diverse fish orders alive, with their venom likely playing a key role in their early survival. For the Scorpaeniformes to make it in the competitive Cretaceous ocean, full of super-sensing, agile enemies like sharks, it wasn't enough to hide. A defensive venom protected them from the diversity of potential predators. But how such painful venoms evolve in the first place still remains one of the greatest mysteries in venom science—we're far more acquainted with how such valuable adaptations are lost.

There's no doubt that the venom of the stonefish or the bullet ant is incredibly effective at deterring predators. But there's a cost to such potent power: venomous animals must constantly produce and store their toxic weaponry. Many venom compounds degrade over time, so animals are constantly producing more venom to replace the old, which is cut apart by enzymes and used for spare parts. That means a lot of energy has to go into making venom, all day, every day. The more an animal must invest in an adaptation, the stronger the selective force must be to maintain it. Understanding why a species became venomous can tell us a lot about evolution—but so, too, does understanding why species lose their toxicity.

Many clades of venomous animals contain an abundance of species. The incredible diversity of bees, wasps, and ants make the Hymenoptera one of the most abundant orders of insects. The Scorpaeniformes are one of the most speciose orders of fish. So it would seem that venom is a successful adaptation, one that leads to rich diversification. But if venom was so fundamental to the Scorpaeniformes' survival, why did the rockfishes and the groupers lose their sting?

The answer requires an understanding of a core concept in evolution: *fitness*. Fitness describes an animal's relative contribution to its species' or population's gene pool. When it comes to evolutionary fitness, it doesn't matter that Kate Gosselin isn't as funny as Cameron Diaz or doesn't have Oprah Winfrey's money and influence: she has eight kids, while the other women have none. Her genes are moving on while the genes of the others have hit a dead end; thus she is the most "fit." In evolution, it's all about reproduction: survival, yes, but only to serve reproduction. It's the number of offspring you leave, and the number they leave, and so on, that really matters. Any trait that an individual has that increases that individual's fitness is passed on, and has the opportunity to rise in frequency in a population, ultimately changing the course of a species' evolution.

It's not always easy to tell which individuals are the most fit. Sometimes it's the ones that are quite literally fittest—the fastest, the strongest, or the biggest. But, depending on the physical and the social environment, strange traits can be advantageous. *Sexual selection*, for example, is infamous for driving the evolution of the worst possible traits for an individual to have. Sexual selection occurs when one sex, often the female, is choosy about selecting a mate. Sure, anyone can *say* they've got the best genes, but being able to feed oneself or outrun predators in spite of a handicap is what evolutionary biologists call an honest signal. So females start to pick the males that succeed in spite of a challenge; they go for the peacocks with the longest tails, which drag and slow them down, or the loudest bullfrog, even though his resounding calls tell predators exactly where to find him and any female that should venture near. It doesn't matter that these males' "adaptations" endanger them: it's worth the risk, because if they mate and produce the most offspring, then that trait will be passed on.

If you could travel back along a venomous lineage, you could theoretically find these individuals who started the line down a venomous path. Their genetic copy machines malfunctioned for only a moment—a mutation in a gene, or an unnecessary carbon copy of one—but it turned out to be a mistake that gave them even the slightest edge. They passed that advantage to their off-spring, and their offspring's offspring, and the useful error spread.

Most venoms are thought to have derived from initially small changes and duplications in immune system genes, particularly in the genes for enzymes that fight off infectious disease or parasites. The same kinds of enzymes that break down bacterial walls can create bioactive lipids that turn neurons on or off; the same pro-teins that rip apart unwanted parasites also tear through a victim's flesh. It's not hard to imagine how, over time, these same enzymes, proteins, and other molecules were co-opted to new uses as the opportunities arose. But just as venom can provide a huge advan-tage, it also comes at an extreme cost to the animal that produces it—and now scientists are beginning to understand just *how* costly it is.

Some studies have directly shown how expensive venom is by testing how much energy an animal must exert, measured through increases in metabolism, to replace what is lost after venom extraction. Such studies look at a measure called resting metabolic rate, which calculates how much energy the animal uses when it's not exerting itself. Resting metabolic rates reflect the energy needed to perform vital bodily functions such as breathing or blood circulation, independent of the calories required to move. We actually calculate the same measure in people: for example, we can evaluate how effective a workout program is by determining whether it's increasing muscle, which uses more resting energy; and we can measure the energetic costs to the individual of actions essential to reproductive fitness, like keeping a fetus alive during pregnancy. When a woman says she has to "eat for two," for example, she's partially right: the added metabolic cost of a baby to a human female is a 21 percent increase in resting metabolic rate.

Similarly, the metabolic cost of making venom is quite steep. One study in snakes found that to replenish their venom supply, the snakes had to raise their resting metabolic rate by 11 percent for three days. Another study found that the death adder, an Australian elapid, raised its resting rate by 21 percent for the first three days of venom production. In other words, during that time period after depletion, about one-tenth to one-fifth of a venomous snake's general energy expenditure is devoted to venom production. In contrast, most studies suggest that adopting an intense exercise routine for several months increases resting metabolic rate only marginally, by less than 10 percent on average. So for the snake, producing venom costs at least as much as an intense workout regime, or even as much as carrying around a baby. And in other species, the cost can be even higher. Scorpions have been shown to increase their resting rates by 20 to 40 percent for up to eight days when replenishing venom. That's a lot of energy to spend on toxins!

It would be expected that if toxins are so expensive, species would be reluctant to use them unless absolutely necessary. Evidence for such *venom metering*, or *venom optimization*, would further support the idea that these animals use their costly venoms only when they feel they have to. Indeed, many studies have found that venomous species do their best not to use venom when it either isn't required or won't be effective. Adult scorpions are not only armed with venom; many also have large, strong, pincer-like appendages called pedipalps that can be used to catch and subdue prey. Scorpions have been shown to prefer to attack with their pedipalps first, resorting to venom only if such an attempt is unsuccessful. Scorpions will also vary the use of their stinger based on the prey: larger prey are more often envenomated than smaller prey, which the scorpion can catch more easily without venom. In some studies, scorpions used their stingers less than one-third of the time. Similarly, when snakes bite offensively, they almost always inject venom. But when they bite people, anywhere from 20 to 50 percent of the bites are dry. Since snakes don't bite people with the intent of eating them, it makes sense not to waste precious venom on us when a bite alone is usually enough to get the message across.

Every individual has a limited energy budget, and to be fit, included in that equation must be the energy needed to find a mate, do the deed, and produce offspring. Your body will decide how to divvy up the calories from that hamburger you just ate into growth, fueling muscles, storing fat for later, and an uncountable number of potential expenses. Without even thinking, we budget how our energy is spent based on the availability of food, mates, and predators. Evolution is a meticulous accountant, penalizing mercilessly for every misspent penny.

Back in the late Cretaceous, more than 65 million years ago, when oceans were dominated by vicious sharks, giant marine reptiles, and other big, toothy species, an ancestral scorpaeniform fish happened to luck into venomous mutations, and its offspring survived and diversified. But as the seas changed, selective pressures changed, and for some of that individual's descendants, venom became an unnecessary expense. There were those that still focused on venomous defenses, such as the stonefishes, lionfishes, and scorpionfishes. But individuals in other species with random mutations that disrupted venom production were able to survive and reproduce as well as their venomous kin, if not better. They lost their sting, though you can still find traces of once-potent proteins in their genomes. The rockfishes and the groupers were born.

While I may have avoided the bullet ant's intense sting (for now . . .), and I have never been stung by a lionfish or its relatives, I don't have a perfect track record when it comes to dodging venomous defenses. And I'm not just referring to the humble bees or wasps that just about everyone has had a run-in with if they've been on this earth for more than a couple of decades. Sadly, I have felt the sting of a serious defensive venom, the kind that even Schmidt would have to rate above a 3.0.

In retrospect, it wasn't the smartest thing I've ever done. Not the dumbest, sure, but probably up there in the top ten. I simply *knew better*. Just a moment before it happened, a voice in my brain screamed that something was wrong, that I needed to stop, that nothing good could come from this. I didn't listen. Instead,

I reached my bare hand into the box and grabbed that damned urchin.

For several years, I have given up my mornings for a week in April to help teach second graders in Hawaii about marine life. Each day, about fifty kids and their parents and teachers go out to these beautiful tide pools off of Māʻili Beach Park in Waianae armed with clear plastic boxes, nets, and reef walkers. For an hour or so the group collects whatever they can find—lots of sea slugs, hermit crabs, nudibranchs, urchins, and brittle stars. Daring parents team up to catch small eels. Every once in a while, they'll find an octopus or a frogfish. Then they bring them to me and my small army of graduate student volunteers on the beach, where we await them with buckets to sort their living treasures. The kids gather together while my student recruits and I tell them cool facts about what they've found. We explain how certain hermit crabs have a mutualistic relationship with the anemones on their shells, or how the collector urchins hide from predators by covering themselves in whatever they can grab onto. Then the kids get the opportunity to touch the different organisms. They *ewwwww* and *awwwww*. Seeing their faces light up when they feel the slime from a sea hare or have their fingertips tickled by a moving brittle star is, hands down, one of my favorite parts of being a biologist.

You would think that with so many kids running around it would be hard to keep things under control, but instead, it's one of the most well-planned field trips I've ever been a part of. For years I have been in awe of how seamlessly and easily the days flow. Every kid has an adult chaperone, usually a parent, who keeps them with the group and behaving well. The teachers are out on the beach and set up long before the kids arrive. All I have to do is show up, bring some friends, and answer questions about marine invertebrates, and everything else is clockwork.

Except, of course, *that* day.

That day, another school group showed up on the beach because they didn't check to make sure it was free. *That* day, there were twenty more kids and parents than usual to begin with, not counting the fifty or so from the other school. *That* day, they had to collect farther out, so everyone brought their finds in at once. I had fewer volunteers than usual, and one of them was my ex-boyfriend, whom I had broken up with less than a week before. And *that* day was the only day anyone has ever found wana.

Wana (pronounced "vah nah") is the Hawaiian name for the Diadematidae family of sea urchins. Around twenty species of urchins are found in Hawaii, most of which are completely harmless. The rock-boring urchins *Echinometra mathaei* and *Echinometra oblonga* are known to locals as 'ina. Though they're difficult to pry out, their stout spines make them perfect show-and-tell organisms for the kids—just sharp enough to seem dangerous, but blunt enough that no one gets hurt. The collector urchins, *Tripneustes gratilla*, or hawa'e, are a local delicacy. There are helmet urchins (*Colobocentrotus atratus*) and pencil urchins (*Heterocentrotus mamillatus*), neither of which has sharp enough spines to stab you no matter how hard you grab them. Then, of course, there are the wana: *Diadema* and *Echinothrix* species, especially *E. calamaris*, the banded urchin, some of which are unfortunately not well banded at all.

To an untrained eye, or, say, a graduate student surrounded by out-of-control second graders and trying to act cool around the ex she's seeing for the first time since the breakup, the black morphs of *E. calamaris* are somewhat similar in appearance to the black boring urchins—except, of course, for all the ways in which they are very, very different. *Echinothrix* have much longer primary spines and a second set of shorter, thinner spines that should really be avoided at all costs. Like other members of the urchin family Diadematidae, banded urchins don't just look menacing, they're also armed with a potent and painful venom—which is why it wasn't the best idea for me to reach into a clear plastic box and grab one.

I was distracted, sure, but I knew that the black urchin in that box was not like the others. It was bigger. The spines were longer and sharper. It just didn't look right. But in that moment I made the decision to try to move it anyway. The kids were restless and surrounding me, I was emotionally on edge, and I needed to clear the animals from the boxes, sit the little ones down, and tell them about what they had found. As soon as I touched it, though, I knew I'd made a huge mistake. Six short spines embedded in my finger, releasing a dark purple fluid that discolored each point of entry. Several four-letter words went through my mind, and I bit down hard to ensure I didn't say them aloud. I knew it was going to be bad.

It *hurt*, but at first, I thought I could handle it, so I braved the pain to continue at my task. My volunteers were still out in the water, and I didn't want to leave the kids without someone managing the beach. Carefully keeping my spine-studded finger elevated, I continued to sort creatures from boxes to buckets. But the throbbing sensation in my finger grew more intense. About ten minutes after the sting, I started to feel lightheaded and nauseated. My chest felt tight. Then, of course, my ex sauntered up and put his arm all too comfortably around my waist like he was still my boyfriend and asked if I was okay. It was all too much. I couldn't breathe. I had to get away, and I needed to get those spines out of me.

Sea urchin stings aren't deadly, though that fact is easy to forget as intense, burning pain radiates from your swollen finger and you struggle to keep from vomiting. I staggered up the beach to the first-aid area, cursing under my breath as soon as the kids were out of earshot. The pain was getting worse, and I was definitely

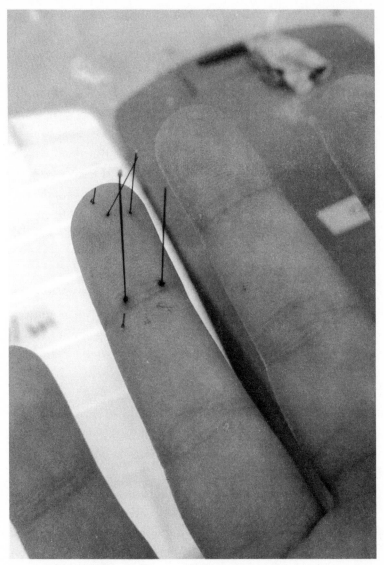

My encounter with the venomous wana

(Photograph by Christie Wilcox)

feeling the systemic effects of a bad sting—according to the medical literature, "dizziness, palpitation, weakness, muscular paralysis, hypotension, bronchospasm, and respiratory distress may occur."

I quickly took stock—no vinegar, no hot water. Damn it. Both of those things would have helped, as heat inactivates the venom components and vinegar dissolves the spines. I desperately rummaged through my first-aid kit to find tweezers. I'd already let the spines stay in for almost half an hour at this point, slowly pouring more and more venom into my flesh. I just needed . . . *Aha! There they are!* I placed the tweezer tips at the base of the first spine and started to tug. Intense pain shot through my finger and I released, but the spine did not give. *I don't know if I can do this,* I thought.

Thankfully, one of the other grad students arrived just in time to take over, and she carefully yanked all six spines out of my hand. Within minutes, the pain began to subside. Less than an hour after the sting, I was able to return to the group and (carefully) help finish out the day. It was my first encounter with the incredible pain of a defensive marine venom. I hope—perhaps naively—that it will be my last.

As a scientist, I wonder why the wana possesses such a potent sting, while the 'ina does not. Although we know some of the compounds that make venomous defenses so painful, and we know why these species possess them, we still don't know how they got them in the first place. Sure, we know the basics—toxins arise when random gene duplications of potential toxin genes lead to copies released from the duties of their regular functions, giving them the freedom to mutate and perform new tasks. It's the same with predatory venoms: duplication frees genes from their prior obligations to the body, making room for novel activities. But what selective forces led to the increase in potency in these toxins, and what maintains their high degree of toxicity?

For snakes and other species that use venom to catch prey, we have a pretty good idea how that works. We know all the players involved. Just as we can look at coevolution in species that have evolved immunity to venoms, we can examine the relationship between predatory venom activities and the bodies of their target victims, and unveil the mysterious hand of evolution that honed venom components to be precise toxins that aid in prey capture or digestion.

But defensive venoms are different. They have to act against a diverse set of possible predators that may have very distinct physiological systems. Bullet ants, for example, might need to defend their nests against mammals, birds, and even reptiles. And their venoms have to act without causing the same intense pain to their own bodies, which means species that employ defensive venoms have to protect themselves from their own universal toxins. It's possible that some do this by isolating their toxins in reinforced compartments—a good strategy, but one that leaves them defenseless if they are envenomated by one of their own. But others are resistant to the toxins themselves, able to shrug off a bite or a sting like it's nothing. For many animals, we still don't know how they pull that off.

Defensive venoms also tend to be simpler in composition—perhaps a consequence of being universal—and they tend to act on our fastest physiological systems: our nerves. Defensive venoms can't afford to work slowly; when a predator attacks, the faster it regrets its decision, the better for the potential meal. If a venomous fish took minutes to induce pain, for example, that would mean minutes of being swallowed and digested (and likely death) before the so-called defense kicked in. Nerves relay signals at unbelievably fast speeds; thus the venom can produce near-instant results. Pain not only teaches the predator a lesson, it's quick.

Sadly, we currently have a very limited understanding of how selection acts on defensive venoms, but there are clues—like the convergence upon pain production—that tell us it must. But to understand how defensive venoms have evolved, more research on these venoms is needed. For understandable reasons, defensive venoms have been less of a priority to scientists. Instead, money and time are focused on species whose venoms cause serious medical issues. While we lack a deep understanding of defensive venoms, we know much more about how animals develop venoms that target other body systems, including those that impact our most vital tissue: our blood.

5

BLEED IT OUT

For the life of the flesh is in the blood.
                                    —LEVITICUS 17:11

When you're going to be traveling in the Peruvian Amazon, there are several essentials to bring with you. Bug spray containing as much DEET as you can stand, to keep the malarial mosquitoes and other disease-carrying biters at bay. Thick socks to protect your feet for long days of hiking over rocky and muddy terrain. Serious rain gear—it is a *rain*forest, after all. And, of course, enough clothes so that you can change and stay comfortable and dry in between forays into the jungle.

For my first week in the Amazon, I had none of these.

When I stopped over in Los Angeles on my way to Peru, I was assured that my checked bag had successfully made it from one airline to another. It hadn't. When I arrived in Lima, I learned that my bag had not left Los Angeles and would take until the next morning to get to me. Seeing as I was flying out to Puerto Maldonado in the morning to head immediately upriver to the Tambopata Research Center in the Tambopata Reserve, I expressed concern over the timing. I was advised to stay an extra night in

Lima and await my lagging luggage, which I did. It still didn't show. So, now assured that my bag would be forwarded to Puerto Maldonado when it showed up (tomorrow, they said), I continued on my journey. It would be another five days before I received my carefully packed clothes and field gear. I entered the jungle in a tank top, skinny jeans, and hiking boots, armed with my camera and computer—the only items I had carried on the plane.

The first few days weren't so bad. I hung my clothes out to dry each night and did my best to stay clean. But by the fourth day, the rains really set in. I cringed every time I got dressed in my still-soaked, muddy clothing. My feet blistered from the lack of adequate cushioning. No amount of soap made me feel clean. By day seven, I resembled the wild pig–like animals in the forest called peccaries: you could detect my presence in the general area by smell alone.

But I wasn't going to let discomfort keep me from finding the animals I had traveled several thousand miles to see. I was in search of one of the deadliest animals in the Amazon: not a jaguar or an anaconda, but a caterpillar capable of making me bleed to death from the inside out.

We humans have a complicated relationship with our blood. The word itself appears some four hundred times in the Bible alone. To the ancient Hebrews, blood was considered the fluid of life. As such, it belonged to the creator; thus the consumption of blood was strictly forbidden (food animals were to be bled dry before their meat could be eaten). The connection between life and blood made it a proper, purifying sacrifice and made even menstrual blood cleansing. And the tribes of Israel weren't alone in consid-ering the red fluid to be of special importance; the practice of spilling blood or consuming it (in a ritual or just for nourishment) has existed in dozens of cultures worldwide. But the so-called essence of life was also commonly blamed for death and disease. For centuries and across cultures, blood has been seen as the cause of sickness, leading to an incredible diversity of bloodletting procedures and implements.

Blood, perhaps not surprisingly, is quite the precious commodity in our bodies. The liquid connective tissue makes up 7 to 8 percent of our body weight, and functions as the major transportation highway between organs and tissues—basically, it's the physiological equivalent of FedEx. It packages and ships *anything*. Your lungs use blood to export oxygen and import carbon dioxide in exchange; the digestive system carefully breaks apart meals into mailable units and ships them via the blood to fuel other organs; and blood takes all of the body's trash and delivers it to the liver and kidneys for removal. Nutrients as well as waste products fly around our body's postal system with remarkable efficiency, ensuring that all organ systems work together and stay in business. Our blood even transports immune cells, serving as the delivery system for internal defensive units.

About 40 to 50 percent of the total volume of blood consists of red cells, which, as the name implies, are what give blood its color. These cells are packed with hemoglobin, an iron-containing molecule that allows for the shuttling of oxygen around the body. Platelets, or thrombocytes, make up a much smaller percentage of blood volume, in part because they're teeny: platelets are only one-fifth the diameter of red blood cells, so while there are anywhere from 150,000 to 350,000 per microliter of blood, they make up less than 10 percent of the total volume. White cells, or leukocytes, are the largest of the blood cells and the key players in the body's immune system, but despite their size and essential job, they constitute only 1 percent of blood's volume. And last, there's the plasma, the slightly yellowish watery solution of sugar, fat, protein, and salt that carries around the colored cells and platelets and makes up about 55 percent of human blood volume.

Red cells carry oxygen, white cells fight off infections, but platelets are tasked with one of the most important jobs: making sure we keep the blood we need to survive. Platelets are our body's wound response team. As soon as a vessel is ruptured and begins to bleed, platelets arrive at the scene and begin to repair the damage by creating what we call clots. They become sticky and clump together with other blood cells to plug the hole, ensuring that precious blood is not lost. Without platelets and the compounds they carry, we would be unable to stop bleeding from even the slightest of wounds. And that's exactly what many hemotoxic venoms and the species wielding them, from moths to mosquitoes, are counting on.

Imagine yourself in Rio Grande do Sul, in southernmost Brazil, merrily going about your day, when all of a sudden it's clear that something is very, very wrong. Your hand starts to swell. You feel dizzy, nauseous, lightheaded. There's a coppery taste in your mouth, like you're rolling pennies around on your tongue. Then it's like a truck has hit you—massive bruises appear throughout your body, though you've taken no blows. Rushed to a hospital, you discover your body is falling apart from the inside. Your blood is flowing unchecked, leaking through vessels and spilling into areas where it shouldn't be. Massive internal hemorrhaging puts you at risk for cerebral hemorrhage or renal failure. You find yourself fighting for survival without a clue as to why.

Lucky you: you've just had a close encounter with a *Lonomia* moth caterpillar, one of the most venomous insects in the world, and the star of the hemileucine moths, which I went to the Amazon to find. Many people don't even notice they've been stung until the symptoms start, but by then it can be too late.

Many of the moths in the subfamily Hemileucinae (family Saturniidae) are largely unremarkable. They're brown. Furry antennae. Mothy-looking. But what the adults lack in luster, their larvae more than make up for with majesty. Hemileucine caterpillars are some of the most beautiful animals on earth. They're often brightly colored, with stunning reds, greens, and blues, but more important, they're adorned with intricate projections that look like carefully pulled glass protruding from every segment, small trees sprouting from their soft bodies. These exquisite structures may resemble hairs, but to assume they feel like them is a potentially fatal mistake. The hemileucine caterpillars are not really "furry"—they're spiky. Each one of the little pointy tips is armed with venom.

These larval moths kill several people a year in Brazil despite the availability of an antivenom, often because the culprit isn't identified until it's too late. Such deaths are agonizing, with the victims succumbing to multiple organ failure hours to days later. Many animals have venoms that target the circulatory system, but the caterpillars in the genus *Lonomia* have perfected the art of bloody messes. Each of the tiny hairlike spines acts like an ampule; the tip breaks off in the victim, allowing the venom to pour in. It's bad enough to be stung by a single individual, but *Lonomia* caterpillars tend to hang out in clusters, which means that, more often than not, people brush up against several at a time. Such a large dose of this potent venom can induce what doctors refer to as a hemorrhagic syndrome, characterized by bleeding from mucous membranes in the nose and eyes, bleeding from scars, and even internal bleeding into the brain.

Oddly enough, the hemorrhagic syndrome induced by *Lonomia obliqua* venom begins as the opposite of unchecked blood flow:

once inside the victim, *Lonomia* venom components make short work of the body's circulatory system by causing an *excess* of clotting. Lopap, a 185-amino-acid prothrombin activator (named, appropriately, for "*Lonomia obliqua* prothrombin activator protein"), flies around in the blood setting off the body's clotting cascade indiscriminately. Meanwhile, Losac (*Lonomia obliqua* Stuart factor activator), which acts like a serine protease (an enzyme that clips proteins) though it has a different structure, activates another arm of the clotting cascade—factor X—leading to even more clots. Combined, the two cause spontaneous clotting in vessels throughout the body, which is referred to medically as "disseminated intravascular coagulation," or DIC. These clots can be deadly on their own, as they can loosen and travel around until they end up getting caught on something and blocking a vessel, leading to a stroke. But more important, Lopap and Losac cause so much clotting that the human body actually runs out of platelets. Without those platelets available to form clots when needed, the envenomated victim bleeds. Uncontrollably. Even though there's no wound to be seen.

The adult *Lonomia* moth is harmless compared to its terrible young. The average moth lives for only a week. In her brief existence as a grown-up, a female *Lonomia* moth must find a partner, mate, and lay her eggs—*up to seventy of them*—which will hatch into deadly caterpillars some two and a half weeks after her death. The caterpillar that emerges is the longest life stage of the animal; the larvae spend three months as venomous caterpillars, feeding on fruits and potentially stinging unsuspecting victims.

Although I searched high and low in my skinny jeans, I didn't find the caterpillars I was looking for. Ironically, the caterpillars weren't the hemotoxically venomous animals I had to keep an eye out for; *Lonomia* moth caterpillars are one of the most striking wielders of hemotoxic venoms—those that target the blood and tissues of their unfortunate victims—but they're not the most common. The everyday masters of the hemotoxic venoms are known for their strange dietary preferences, and have inspired entire genres of horror movies and dramatic rom-coms: the vampires.

What we call vampires are what scientists call hematophagous animals. Hematophagy, or blood-eating, is one of the most specialized dietary adaptations on earth, and consequently, the parasites that dine on blood are among the most specialized of all venomous fauna. And, yes—all vampires are venomous. Every species that drinks blood produces a special toxic slurry necessary for slurping the precious bodily fluid from its host.

It is one thing to kill your prey, but quite another to get close to it, taste its precious lifeblood, and escape unnoticed. Hematophagous animals need venom not only to access their food source but also to fool it, lest they end up squashed under a quick-tempered hand. Thus the venoms of mosquitoes, ticks, and even vampire bats are all strikingly similar. Nature's most precise phlebotomists are armed with venoms that include pain-killing substances to mask where the blood is flowing from, compounds that either fight off the host's immune defenses or enable the vampires to cloak their blood-removal devices from immune surveillance, and anticoagulants to disarm and disrupt blood-stopping clots.

Whenever I think about blood-feeding animals, I am taken back to 2003 and some of my fondest memories from college. When I was a freshman, one of the required courses for my undergraduate degree was Invertebrate Biology. My professor was known as a hard-ass when it came to grading, but as a parasitologist, she had the most fascinating stories (it's a simple fact: parasitologists *always* have the best stories). I'll never forget how she started the lecture on the subclass Hirudinea in the phylum Annelida, otherwise known as the leeches. She had owned pet leeches, she told us—Dracu and Dracu II—and she used to bring them in for show-and-tell. She doesn't anymore, she explained. From what I remember, this is why:

One year, she brought Dracu to her class to show her students what a living leech looked like. Excitedly, she carried her beloved pet to the front of the room in a clear container filled with water, and gently coaxed him to swim ("him" was an arbitrary choice, as leeches are hermaphroditic). Though they are considered gross and disgusting because of their diet, leeches are quite beautiful swimmers: they undulate in sinusoidal perfection to move from one area to another. After showing off Dracu's aquatic acrobatics and explaining the musculature involved, she then calmly placed him on her arm and allowed him to feed while she explained his mouthparts and the anticoagulants in his venom. She always fed her pets herself (I've found this is true of many parasitologists). Like most leeches, he would usually suck her blood for a few minutes, growing in size as he downed his warm, delicious treat, until he was finally content. Then he'd simply fall off, sated.

One day, though, Dracu happened to hit a really good vein. When he was finished, he let go as always, but she kept bleeding. And bleeding.

My professor had underestimated the power of Dracu's anticoagulant venom this time, and soon found herself desperately trying to sop up a growing pool of blood in front of her entire class of students. Apparently, the incident upset some of those in the classroom, and she was politely told that she probably shouldn't have leeches feed from her in her lectures anymore.

Dracu, like other real-life vampires, is equipped with a venom that aims to do exactly what it did: cause unrestricted blood flow. As soon as a leech (or a mosquito, or a vampire bat) punctures the flesh of its dinner host, the blood starts to thicken. Platelets begin the coagulation cascade when they interact with adhesive molecules such as collagens and fibronectin that exist in the complex gel-like substance between cells, termed the *extracellular matrix*, or ECM, which is exposed when an animal is wounded.

Platelets stick to the ECM using a series of interactions between platelet constituents and ECM protein parts, including platelet receptor glycoprotein binding to von Willebrand factor (vWF) and collagen receptors binding to collagens. The binding of collagen spurs the release of thromboxane A2 (TXA2) and adenosine diphosphate (ADP), which activate platelets and spur pathways that lead to platelet aggregation, bringing more and more platelets into the area. They're assisted by the release of epinephrine and serotonin from the activated platelets, which further induce changes that promote aggregation. Activated platelets also promote the generation of thrombin, one of the most potent compounds when it comes to blood coagulation. The end result is a big, sticky mass that plugs the hole: a clot. Caterpillar venom aims at inducing clots, but vampires have to prevent or bust them at all costs.

It turns out it's quite difficult to drink clotting blood, much like trying to suck a chunk of banana in your smoothie through the straw. So vampiric animals' venoms—injected along with their saliva when they bite—contain anticoagulants to prevent such blockages. These bloodsucking species don't rely on a single venom compound to keep their meal flowing. Instead, each is armed with many different anticoagulants. They range in molecular size from a wee five kilodaltons, or kDa (weighing only a few billionths of a trillionth of a gram!) to thousandsfold larger. Scientists have described the diversity of clotting inhibitors found in venomous vampires as "remarkable," as there are toxins to target every step of our coagulation cascade.

Some venom molecules start at the beginning of the clotting cascade, binding to the platelet receptors or exposed ECM components such as collagen. Others break down or tie up ADP, TXA2, epinephrine, and serotonin to keep them from acting. Then there

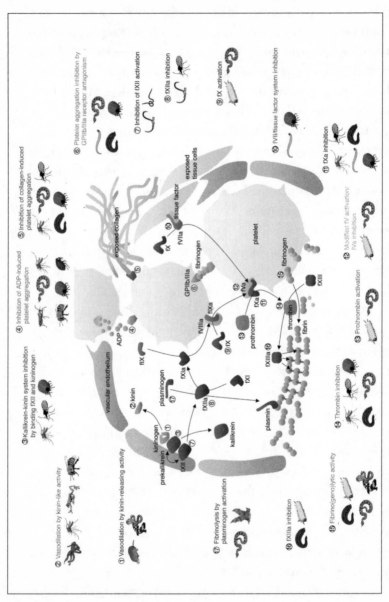

The diversity of hemotoxic venom compounds that disrupt the coagulation cascade

(Figure © Bryan Grieg Fry)

are the ones that act further down the line, blocking thrombin and its key role in coagulation. There are enzymes: phospholipases, metalloproteases, hyaluronidases, and apyrases. There are lectins, lipocalins, and peptidases. Varieties of these diverse compounds are found in all vampiric venoms, from the much-maligned mosquitoes to fleas, ticks, and vampire bats. And they are found in abundance in leeches like Dracu. Dozens of anticoagulant compounds have been isolated by venom scientists from leeches alone.

Indeed, leeches—and their anticoagulants—have been used medicinally for centuries. Long before modern medical science could explain that there are more than sixty different bioactive compounds in leech venom, leeches were used to treat a variety of diseases. Even today, doctors still apply live leeches to patients; they can help improve circulation and decrease your body's chances of rejecting new tissues during grafts and reattachments, and they are even used by clinicians to treat varicose veins. In fact, an anticoagulant used during modern surgeries like angioplasty—Angiomax (Bivalirudin)—is a small peptide based on a venom compound from *Hirudo medicinalis*, the medicinal leech. And that's just one of three venom-derived anticoagulant pharmaceuticals currently marketed in the United States. That means anticoagulants account for *half* of the six venom-based drugs that are FDA approved—and there are several more toxins transformed into treatments in clinical trials. The ability to regulate the circulatory system is hugely important for doctors because it's always a medical emergency if something goes wrong with it. The fact that hemotoxic venom components can manipulate heart rate, blood pressure, and coagulation is exactly what makes them such excellent candidates for pharmaceuticals—which you'll learn a lot more about a little later on in this book.

Almost as important to the vampires as the anticoagulants are the other components found in their venoms. Most species that are hematophagous feed on much, much larger organisms. To do that, they need to slip in unnoticed, which means neither the conscious nor the unconscious defenses of the host may be alerted or triggered.

To keep their meal ticket unaware of its unwitting donation of blood, vampiric venoms contain analgesics and anti-inflammatories to dull pain, as well as components to dismantle or quiet immune responses in the host. Some of the anticoagulant compounds perform double duty, as pro-clot compounds like serotonin and thrombin also trigger pain and inflammation pathways, while others are dedicated to their individual tasks to prevent detection. Ticks, for example, possess a diversity of painkilling and anti-immune compounds because they may stay embedded in their host for days at a time; they face a much greater risk of discovery than the flitting mosquito, which can be in and out in minutes.

It's the incredible synergy of dozens of venom compounds that makes vampiric venoms so astonishing. Their venoms swiftly and effortlessly dismantle one of our most vital body systems without causing serious harm—at least until hitchhikers get involved: the pathogens of insect-vectored diseases. If it weren't for the species that can ride on vampiric venoms to make their way into their victims, including parasites such as malaria and dengue and other viruses, hematophagous animals would be basically harmless. Vampiric venoms are as precise as they are benign. If anything, the less damage done to the host, the better for the vampire: the healthier the host is after being fed on, the more likely it is to provide another meal down the road.

Of course, not all species that wield hemotoxic venoms are as delicate with their delivery as leeches and other vampires. Nor are their wounds as unnoticeable as the caterpillars'. The largest species possessing a hemotoxic venom is not known for its subtlety. I'm talking about one of the most legendary reptiles alive: the Komodo dragon.

I've wanted to see a Komodo dragon in the wild for just about as long as I can remember because of my favorite book, *Last Chance to See*. It's written by Douglas Adams—yes, *that* Douglas Adams, the one who is best known for his science fiction novels involving interplanetary hitchhiking. Don't get me wrong—*The Hitchhiker's Guide to the Galaxy* and the rest of his books are marvelous. But in *Last Chance to See*, Adams managed to pack all the concepts that make his fiction great into a nonfiction book about conservation. Adams simply tells stories of traveling around the world with the biologist Mark Carwardine to see some of the most endangered species on earth, effortlessly recounting his experiences with creatures from bureaucrats and pickpockets to finicky parrots and man-eating lizards.

It was in *Last Chance to See* that I first read about Komodo dragons. The creatures some claim led to the cartographic warning "Here Be Dragons" are among the most notorious animals on the planet. The Komodo dragon (*Varanus komodoensis*) is the largest living lizard in the world. The biggest one on record measured more than ten feet long and weighed more than 360 pounds. They're intimidating beasts known to feed on anything they want, from pigs to water buffalo (which can stand more than six feet tall at the shoulder and weigh more than one thousand pounds). Komodos' massive jaws are packed with inch-long teeth that tear through flesh with ease. They sense their prey much as a snake does, with a forked tongue that accurately detects the smallest concentrations of odors in the air. And according to many scientists who wrote about them in the 1980s (and Adams, in his chapter on the dragons), the worst part of these deadly beasts is that their saliva is fetid with deadly bacteria from the decaying flesh they feed on. If prey don't fall victim to the wounds at first, the sepsis-inducing bacteria will finish them off. But they, and Adams, were wrong.

For a long time, civilized scientists thought that the rumors of dragons in Indonesia were stories made up by sailors. It wasn't until the early twentieth century, when full specimens were collected, that the tales of old were finally believed. Now there is no question: the dragons are real. But modern science has also shown that their legendary bacterial bite is no more than a fairy tale.

People didn't just make up the idea of a bacterial bite out of nowhere. The notion that Komodo dragons harbor virulent bacteria in their mouths was based upon observations made by the scientist Walter Auffenberg in the 1970s. He noticed that dragons would attack water buffalo, but were rarely successful in killing them in that initial attack. The buffalo would flee, but the dragons wouldn't just move on to a new food item—they'd follow the animal for days at a time. The buffalo wounds soon became gangrenous. Eventually, the massive beast would die from the infections or be so weakened that it no longer could fight off the oversized lizards. Then the dragons would feed.

This led Auffenberg to suggest that "induction of wound sepsis and bacteremia through the bite of the Komodo dragon may be a mechanism for prey debilitation and mortality." Though the idea seemed far-fetched, scientists isolated pathogenic bacteria from Komodo saliva, supporting the crazy idea. And in some ways, it's not that hard to believe: many species use bacteria to produce deadly toxins. The blue-ringed octopus, for example, doesn't make its own tetrodotoxin—a bacterial symbiont that lives inside the animal's tissues does the hard work for it. There are plenty of species that steal poisons from their diet, and some nudibranchs (marine slugs) can even steal whole stinging cells from the anemones and jellyfish they eat and use them for their own defense. But the idea of a living venom of sorts—culturing virulent bacteria to take down prey—is entirely unheard of.

Bryan Fry, the venom expert at the University of Queensland, simply didn't buy the whole idea of a bacterial bite. He'd seen firsthand that Komodo dragons are clean animals. "After they are done feeding, they will spend ten to fifteen minutes lip-licking and rubbing their head in the leaves to clean their mouth . . . Unlike people have been led to believe, they do not have chunks of rotting flesh from their meals on their teeth, cultivating bacteria." Furthermore, the facts didn't add up. Water buffalo are a newcomer to the islands of dragons—Komodo dragons evolved eating smaller prey, the size of pigs or little deer. The dragons don't need bacteria to take those animals down—when they bite animals of that size, the prey bleed out in less than an hour.

In addition, Komodos, like other monitor lizards (large lizards in the genus *Varanus*), are close relatives of snakes. In 2005, Bryan and his colleagues showed that all the species in this lineage of reptiles share the same venom genes, which implies that the early ancestor of all monitor lizards and snakes was venomous. But those who insisted that dragons possessed a bacterial bite were unimpressed. A few years later, Bryan and his team scanned the head of a Komodo dragon with an MRI machine—the same tech used to look for injuries in our bodies—to show that Komodos do indeed have venom glands, and their lab studies demonstrated that the toxins they produced could cause severe drops in blood pressure. Those unable to let the myth go insisted such findings were "meaningless, irrelevant, incorrect or falsely misleading."

Finally, Bryan performed the ultimate test: he and his team repeated the previous research that had cultured bacteria from the mouths of dragons, this time with more samples and better techniques. They didn't find the pathogenic species that people claimed were living there. Instead, the oral flora of the dragons was just like any other carnivore's. The overall finding was clear: rather than sepsis waiting to happen, the bacterial community in Komodo spit "is reflective of the skin and gut flora of their recent meals as well as environmental flora."

But if the dragons aren't introducing deadly bacteria into the animals through their bites, why do the water buffalo get infected so often? The answer is in their name. In the freshwater marshes they are native to, water buffalo splash around in clean, flowing rivers and cool, clear pools. But there is very little clean water on Rinca, Komodo, or any of the other islands where the dragons roam, because they don't have mountains to feed streams, or extensive aquifers of clear, fresh water. Instead, the buffalo wallow in the only water they can find: warm, feces-filled pits. While smaller prey bleed out quickly from the trauma caused by the Komodo's teeth and the hemorrhagic venom they deliver, the water buffalo tend to get away with seemingly innocuous wounds. But then they do what water buffalo do best, and incubate their wounds in bacteria-filled cesspools, exposing themselves to the very pathogenic bacteria that scientists once believed lived in the dragons' mouths. "Having gotten septicemia in Flores from deep lacerations resulting from a boating mishap," Bryan noted, "I can attest to how quickly such environmental sources can produce life-threatening infections." Even the previous study's finding of virulent bacteria in 5 percent of dragon mouths makes sense: those dragons likely had recently drunk the same disgusting water.

In the end, it is the buffalo's love of water that kills them, not the dragon's venom. But while the dragons' mouths aren't rife with pathogenic bacteria, they are still quite venomous. For species smaller than the buffalo—including us—their devastating bite combined with potent venom is more than enough to kill.

Komodo dragons don't have fangs like snakes. Instead, they have an extremely complex reptile venom gland: venom is stored in five glandular compartments in their bottom jaw, and travels through separate ducts to be released in between serrated teeth. In total, each lizard can store more than one milliliter (1 ml) of venom in its mandible compartments (that's a little more than one-fifth of a teaspoon). The venom itself comprises thousands of components. It attacks the cardiovascular systems of mammals with a vengeance, causing blood pressure to plummet, inhibiting coagulation, and inducing shock. Included in the toxic mix are kallikreins, which release vasodilators that widen veins and arteries, leading to deadly hypotension and type III phospholipase A_2s, which act as potent anticoagulants. Noticeably lacking are the kind of neurotoxins that some of its relatives are known for— but that makes sense, when you think about the dragon's end goal. Unlike many snakes, the dragon isn't going for paralysis; it's aiming for exsanguination.

When you combine the Komodo's terrifying teeth with its potent venom, you have a huge lizard that can kill with frightening efficiency. You might think that a mouth full of inch-long serrated blades would be enough—and it often is—but if the bite itself is insufficient, the venom finishes the job. The maw full of razor-sharp teeth creates gaping wounds that bleed profusely, but even an otherwise nonlethal bite from one of these dragons becomes deadly as the dragon's venom ensures that blood keeps flowing and flowing until the prey's blood pressure tanks, inducing shock. If the blood loss doesn't kill the unfortunate victim, the shock does—and the dragon can claim its meal.

I finally got my chance to see these venomous beasts on a research trip to Bali. We arrived in Labuan Bajo on the island of Flores in the late afternoon. It was just me and Jake—Jacob Buehler—the man I had fallen completely in love with, if not for his atrocious punnery then for his complete lack of hesitation when I asked if he wanted to join me on a mission to Indonesia to find dragons just a couple of months after we started dating. I would come to realize that this was perhaps the most spontaneous thing he's ever done and slightly out of character. Unlike me, he prefers to think things through from every angle before deciding to jump in. But it's a good thing he came along; he is far more cautious than I am, and on more than one occasion, his prudence has proven superior to my carefree attitude. After all, when we were in Ubud, on Bali, he was the one who told me not to buy the bananas to feed the monkeys at the sacred forest ("I've heard they bite, Christie"). I didn't listen to him, of course ("I want a monkey on my shoulder, Jake!"), and was promptly bitten in the leg by a grumpy, foot-tall macaque who decided that the banana I offered wasn't good enough. I had a massive bruise for more than a week, and the wound required eight shots of immunoglobin and four rabies vaccinations. But hey, I got my monkey picture! In several other cases, Jake's more reserved nature has saved me, and when you're traveling to an island known for its massive man-eating lizards, you really can't be too cautious.

We booked the flights last-minute. The plane was small and shaky, and I couldn't help but wonder if it would make it even the short jump from Denpasar, Bali's capital, to Labuan Bajo. And then there was the boat we hired for the day to take us to Rinca Island, where the dragons live—a rickety wooden vessel with an engine that took four tries to start, captained by a man with a 1970s pornstache who spoke less than a dozen words of English. Not to mention, of course, that the whole point of the adventure was to get up close to a legendary killer.

It's not hyperbole to call Komodo dragons man-eaters, for they have been known to eat people should the opportunity arise. In 2008, a group of lost divers barely survived two sleepless nights stranded on Rinca—they had to fight off large dragons by swinging their dive weights at them and throwing rocks at their heads. There are enough people who have not been so fortunate.

As we neared Rinca, it became quite clear that this was not a land to be trifled with. Jagged cliffs rose ominously from the water. It was the dry season, and dead brown grass covered the hillsides, with sparsely distributed scraggly trees jutting out of the arid earth. Rinca stood in such stark contrast to the wet, vibrant rainforests on Bali. Everything about the island seemed rugged and dangerous, and I imagined how foreboding it must have appeared to early travelers landing here for the first time. And if the island itself wasn't daunting enough, when they finally came ashore, they would have learned the hard way of the beasts that call Rinca home. I half expected to see an old wooden sign hanging at the docks warning away potential trespassers. *Beware*, it would say in letters faded by salt and sun. *Here be dragons.*

Our guide on the island was a young, gangly Indonesian man named Akbar. He carried a large stick as he walked, which he soon explained wasn't to help his hiking; it was to fend off dragons or anything else, should the need arise. The dragons aren't the only danger on Rinca; Komodo National Park has one of the highest densities of venomous snakes anywhere on earth, including a dozen deadly species. Even the pigs, the deer, and the water buffalo that are preyed upon by the dragons can be dangerous when spooked. But no need to worry, Akbar assured us. He had a stick.

A short way into the hike, we came upon several skulls that the guides had tied up in a tree. Akbar explained that the skulls were found on similar walks through the landscape, and belonged to animals the dragons had eaten. A young British woman with us seemed particularly unnerved by the sight.

"You don't feed the dragons?" she asked cautiously.

"No, they feed themselves," Akbar answered with confidence.

"What if they don't catch anything? Won't they be hungry?"

The question caught Akbar off guard. He thought about it for a second before replying: "Yes, I'd think so."

"But if they're hungry, aren't they more dangerous?"

Akbar smiled. "Sometimes."

She paused, and then looked at the stick in his hands. "How often do you have to use the stick?"

His grin widened. "Sometimes."

Before long, we spotted our first water buffalo. The humongous animal was peacefully chewing some grass no more than fifty feet off the path. As a group, we quietly walked within about

Here be dragons. (Photograph by Christie Wilcox)

twenty feet of it. I simply couldn't imagine a dragon taking down something so huge. It wasn't just about size—though the beast was massive, it was also well armed, with large horns and sturdy hooves that could crush bone all too easily. I continued forward, camera in hand, to get a better shot, but Jake's hand on my shoulder stopped me. I knew why; though the animal was calm now, if it was frightened or startled, it could easily charge me, and I would be hard-pressed to get out of its way in time. I rolled my eyes, but didn't get any closer. Just a little way farther, our guide found another one wallowing in a small pool of standing water, demonstrating the exact behavior that led to the whole "bacterial bite" nonsense in the first place.

But, of course, I was there to see dragons, not buffalo—and Rinca didn't disappoint. "They don't *look* that terrifying," Jake whispered to me as we stood just inside the entrance to the park. Five large Komodo dragons had decided to nap in the shade of one of the buildings nearby, and a small crowd had gathered around them. He was right. Lying there lazily, the beasts looked slow and cumbersome. We stood no more than twenty-five feet from the closest one, which seemed completely indifferent to our presence. We were staring at a couple thousand pounds of one of the most deadly lizards on earth, and they didn't even seem to notice we were there.

Behind them, a juvenile Komodo ambled about, its long forked tongue flicking. I tried to imagine the bigger beasts in a similar state, wandering around in search of food. I couldn't. *These are fat, happy lizards,* I thought, *that have grown accustomed to foot traffic. I bet Akbar lied about feeding them. These animals don't look like they could snag a pig if their life depended on it. Besides, with all these people around, if they were hungry, they'd have moved by now.* I took a step forward, bending down on one knee to get a close-up shot of the sleepy face of one of them.

Jake coughed gently.

"Come on," I said, "they're not going to move. Just a few steps closer?"

His only response was a stern look.

"Fine," I sighed, and turned away. He was right, of course. Again. Truth is, I'd been bitten by enough wild animals on that trip (including that monkey and a very aggressive fish)—you would think I'd have learned not to push such boundaries. It's better to know the dragon's venom academically and not from personal experience, I reminded myself.

We backed away.

Reflecting on the animals I've talked about in this chapter, one thing is clear: an eclectic set of animals employ hemotoxic venoms. There's such variety in hemotoxic venom wielders because blood-targeting venoms can be used in incredibly diverse

Like I said. Lazy Komodo dragons. Nothing to worry about . . .
(Photograph by Christie Wilcox)

ways. *Lonomia* caterpillars and their kin turn clotting agents into deadly defenses. Komodo dragons use anticoagulants in their venom to bleed their victims dry. Mosquitoes and other vampires have venoms similar to those of Komodos, but use them for small, vampiric tastes rather than exsanguination. And there are many snakes that use hemorrhagic effects to immobilize and kill their prey. In all cases, it's not about individual toxins—it's about synergy. The individual components of these venoms are often not that deadly. But by combining several components working together in perfect lockstep, hemorrhagic venoms become biological machines, able to precisely and efficiently do the bidding of the venom's maker.

Take the Terciopelo viper (sometimes called the fer-de-lance), *Bothrops asper.* It's considered one of the most dangerous snakes in the world. Its broad, flattened head is easily recognized in the snake's homeland of Central and northern South America, as this species is responsible for the majority of venomous bites in that region. These snakes don't possess a single, killer toxin; their venom is made of many components that work together to ravage a victim's circulatory system. When the snake bites, several P-III metalloproteases (enzymes that use metals as a part of their protein-chopping machinery) start cutting up the membranes that anchor capillaries, causing them to break apart and become unstable. Meanwhile, a platelet-aggregating compound called aspercetin starts rounding up the blood components necessary to form clots, leaving the victim with no way to stop the flow. Phospholipases and serine proteases also go to work, adding to the hemorrhagic effect. The snake's target quickly loses the ability to control its own blood and cannot maintain muscle or brain oxygenation, rendering it effectively paralyzed, if not dead from cardiovascular collapse.

When scientists sought to tease apart the various components of *B. asper* venom, they learned a curious thing: the sum of the parts did not equal the whole. In 1993, scientists isolated three hemorrhagic factors, referred to as BaH1, BH2, and BH3, from the venom. As a set, they contained more than half of the venom's overall hemorrhagic activity. But when they were separated, their combined activity was only about half of what it was when they were together. And it wasn't just *B. asper.* Scientists have found synergistic effects of venom toxins in other species of snakes, as well as in bees and hornets.

The synergistic effects of venom compounds might begin to explain one of the biggest mysteries in venom science: why there are so many different compounds in venoms. Upon first glance, it seems quite silly. Why have hundreds of toxic compounds in a venom when one will do the trick? For many of the paralytic venoms, one toxin is enough. So why do the *Lonomia* caterpillars need so many different toxins when one of their compounds is enough to cause massive hemorrhaging? Or why do mosquitoes and leeches make dozens of anticoagulants—surely one or two would be enough? But single toxins can also be a dangerous investment, as the targeted species might develop resistance to that compound, rendering them immune. The caterpillar would quickly become easy pickings if any potential predator were unperturbed by its potent venom, just as the vampires would starve if their venoms couldn't keep blood flowing from their victims. Single toxins only do one thing, while multiple toxins have the potential for modulation and versatility. Understanding the evolution of such multifaceted venoms gives us a better understanding of the selective forces that transform your average animal into a venomous one.

In reality, mosquitoes, leeches, and vampire bats are among the few hemotoxic venomous species whose venom solely targets blood.

Most of the species whose venoms contain anticoagulants, clotting disruptors, and other hemotoxic components, like *B. asper*, also have much more . . . uh . . . repulsive, visceral effects, which I'll discuss in the next chapter. You might want to wait a few hours after dinner before continuing.

6

ALL THE BETTER TO EAT YOU WITH

The poison of her teeth is the necessary means of digesting her food, and at the same time is certain destruction to her enemies.

—BENJAMIN FRANKLIN

The first time I heard a rattlesnake's warning was unforgettable. The buzzing sound of an angry rattlesnake is instantly recognizable, even if you've never heard it before, and the vibrating noise has an unsettling way of shaking your very core, instantly instilling a nauseating feeling of fear. I froze on instinct, unable to determine the direction or distance of the haunting sound. It was so . . . *loud*. Frantically, I looked at my shoes, half expecting to see a snake coiled between them.

"Watch your feet," I recalled Chip Cochran saying just fifteen minutes earlier as we started to climb the scrubby hill behind his house in Loma Linda, California. "Make sure you don't step on a rattlesnake."

I'd met Chip a couple of years earlier at an International Society on Toxinology meeting. (Yes, that is toxi*n*ology, with an *n*, not toxicology, with a *c*. Toxicology is the study of poisons and their effects, while toxinology is the study of all bacterial, plant, and animal toxins.) He didn't exactly fit my mental image of a snake wrangler; I tend to imagine them all as big, burly men with thick, rough skin that even the sharpest fangs can't penetrate. Chip, on the other hand, was just a little taller than me, and with his short blond hair, dimples, and bright blue eyes, he exuded boyish charm. Over a beer on a hotel balcony, Chip cheerfully explained his project on venom variation in speckled rattlesnakes to me and two other grad students. He went on to recount all the other snakes he's found in his "herping" career. His eyes lit up with mischievous joy every time he spoke of an encounter he's had, like when a mamba struck *this close* to his face. How exciting it must be to work with venomous snakes, I thought, so when he told me I should pay him and his lab at Loma Linda University a visit to see what it was like, I happily followed up on the offer. Soon enough, there I was: in the desert to the east of LA, tagging along with Chip; his advisor, Bill Hayes—a renowned herpetologist—and his labmates as they went in search of venomous snakes. Venomous snakes that, apparently, are easy to overlook. It was the third time that week I'd been warned about stepping on them.

The rattle continued as I stood paralyzed—it was eerie and high-pitched. Slowly, I was able to pinpoint the source: a large rock formation to my right. Chip was much quicker, and was already peering into the formation's crevices. "There she is," he said confidently, and waved me closer. In the very back I could see a small rattlesnake curled up tight with her rattle in the air. The rocks had served as a natural speaker, amplifying her sound so that she seemed much larger. In reality, she was only about two feet long, and was at least four feet away, tucked into the cavern—a safe distance. I felt my blood pressure and heart rate drop as I assured myself that the snake posed no immediate threat.

I remembered that rattlesnake bites are a common occurrence

in the United States, but rarely result in death. In fact, out of the eight thousand venomous snakebites that occur annually in the United States—most of which are attributed to rattlesnakes—fewer than a dozen are fatal. Like those of many other pit vipers, rattlesnake venoms are mostly hemotoxic—they target tissues and blood—rather than neurotoxic (targeting nerves). While people often talk about hemo- and neurotoxicity as a binary, venoms don't fall into one category or the other; they exist on a continuum with differing levels of hemo- or neurotoxicity. The deadliest venoms are the ones that are nearly or purely neurotoxic because they cause paralysis, through either nerve signal blockage or overstimulation, particularly of life-and-death muscles of the diaphragm, chest wall, and heart. The main effects of more-hemotoxic venoms, on the other hand, are more gruesome—hemorrhaging and necrosis—but less lethal.

Necrosis is defined as the death of living tissue, but the clinical definition fails to encapsulate just how disgusting and horrible tissue death is. Necrotic venoms leave large areas of skin and even entire limbs rotten and gangrenous, oozing blood and pus and stinking of decay. Healthy, pink tissue becomes black in death, swelling with fluid from liquefied flesh, until it falls from the bone in putrid, zombified chunks. It is no wonder that doctors and scientists prefer the term *necrosis* to a detailed description of such wounds.

Of course, the venom is just doing what its wielder asked of it. Hemotoxic venoms destroy flesh in part because that's what the venom is aiming to do: it helps to have venom that gets the process of digestion started, and a special subset of compounds and enzymes help accomplish this feat. Rattlesnakes use hemotoxins to subdue their prey, but also to begin the long process that turns fur and bone into food. Other hemotoxic species use venom to liquefy their victims, slurping up the nutritious slurry left after their venom does its work. Unfortunately, when these animals bite us defensively, these digestive venom compounds tear through our tissues, leading to pain, swelling, and necrosis.

Rattlesnakes belong to Viperidae—the vipers. In the last chapter, I talked a bit about one of their relatives, the Terciopelo viper, and its hemotoxic venom chock-full of potent toxins. The various venom elements work synergistically to cause complete cardiovascular collapse in the snake's intended prey. But, of course, humans aren't the snakes' intended prey; they bite us only in defense. And since we're much larger, death is not immediate. Of the vipers, the Terciopelo viper and its close cousins in the genus *Bothrops* in particular are known for causing devastating necrosis.

Part of the problem is that these vipers coexist with people in some of the poorest places on earth, where doctors—let alone hospitals stocked with antivenom—are few and far between. In rural South America, Africa, and India, many snakebite victims will receive minimal treatment, if any. A small bite on the leg or hand quickly becomes a festering sore. It's only weeks later, when the necrosis has completely taken over much or all of a limb, that victims are properly hospitalized and begin to receive the care they needed from day one. The popular press has been known to callously refer to what remains of a bitten limb in such cases as "black sticks." The term is as self-explanatory as it is crude, and honestly, the images are nauseating even to a desensitized biologist like me.

Even with antivenom treatment, snakebite necrosis can be severe. Antivenoms work by binding circulating venom compounds in our blood to prevent further pathologies, but they can do nothing for areas where the venoms have already done damage. Hemotoxic venoms are quick-acting and cause severe local damage; thus the antivenom can only help ensure damage doesn't descend into systemic trauma and death. To add insult to injury, scientists have also found that some of the necrotic venom compounds are not immunogenic, which means they slip by the immune systems of the antivenom-producing animals, so there are no antibodies to inhibit them in the antivenom to begin with. And the worst necrotic venoms don't just tear through cells on their own: they enlist our own immune system to continue the death and destruction. Antivenoms are powerless against that.

Necrotic snake venoms start working the moment they're injected by the snake's fangs. Metalloproteases lead the attack, breaking apart structurally important components of blood vessels and tissues, including the critical adherence proteins that keep the cells of blood vessel walls connected into a blood-proof barrier. Local edema occurs quickly as capillaries start to hemorrhage, and the area swells with fluid. The proteases continue the assault on the tissue by aiding in the execution of skeletal muscle, though the exact mechanisms aren't well understood. Not to be outdone, phospholipases launch an assault on muscle cell membranes, ultimately leading to myonecrosis—the death of muscle tissue. Some of the phospholipases punch holes in membranes with their enzymatic activity, cutting apart the phospholipids that form the membrane wall, while others seem to be just as myotoxic without cleaving lipids, by mechanisms yet unknown. Additional venom enzymes, including hyaluronidases and serine proteases, add to the carnage. Meanwhile, while the war rages at the site of envenomation, other venom compounds may escape the action and travel throughout the body, widening blood vessels to cause quick drops in blood pressure that can lead to shock and even death, or causing rapid system-wide skeletal-muscle death, or *rhabdomyolysis*, which releases large quantities of the muscle protein myoglobin, blocking kidney tubules and leading to possibly lethal renal failure.

And that's only the beginning. The venom proteins don't

just do their own damage; they expertly trick our own cells into fighting with them. The massive cell death and certain venom activities, notably the liberation of tumor necrosis factor by metalloproteases and the release of bioactive lipids by phospholipases, activate immune cells that rush to the wound. Our immune cells are trained to fight to the death—which, in the case of bacterial or viral infection, is a great thing. But in the case of snake venom, there's no one to kill. The venom compounds are lone protein warriors, not a cohesive invasive force, but the body's army can't tell the difference. Leukocytes and other immune cells start ramping up inflammatory pathways, producing and releasing cytokines such as interleukin-6 (one of the immune system's messengers), which further signal for an immune onslaught. But with no bacteria to lyse (or other foreign bodies to attack), our body's weaponry has no enemy to target. The immune system thinks it's stamping out invasive forces. Instead, valiant volleys act as friendly fire, adding to the death toll of innocent tissues.

Just how much of the massive necrosis seen in such bites can be blamed on our own immune reaction to the venom isn't entirely clear, but studies suggest it's a lot more than one might think. Scientists have discovered that when the body's own inflammatory pathways are shut down, necrosis from snake venoms is greatly reduced. Any pharmaceuticals that combat our body's response seem to lessen the blow. For example, venom phospholipases can cause a particular kind of immune cell, mast cells, to release histamine—the same stuff that causes strong local and systemic reactions in the case of allergies. But the swelling and edema associated with envenomation can be lessened with simple, over-the-counter Benadryl. Such results suggest that while antivenom therapy is important, immune-quieting drugs may help with the devastating necrosis that antivenoms can't manage—which is especially good news for doctors in rural areas with limited access to antivenoms. However, research on treatment options other than antivenoms has been slow and is in desperate need of adequate funding.

Meanwhile, the venomous vipers continue to disable victims through the ravages of necrosis. *Bothrops* species are responsible for some of the most necrotic bites, but they're not alone in causing devastating tissue death. Necrotic venoms are found littered throughout venomous groups. Though the elapid snakes are generally considered neurotoxic, some of the cobras, like the spitting cobras, can also cause terrible tissue damage. Jellyfish, especially the potentially lethal box jellies, also can cause serious skin damage. And then there are species whose venoms are not generally necrotic, but on occasion can be extremely destructive. Stingrays and even wasps and their kin sometimes cause large lesions. Scientists are still trying to understand what triggers these rarer cases, but clues are turning up in studies of venom components.

The rattlesnake curled in the rocks wasn't the only rattlesnake I got to see while I visited Chip. Loma Linda University is home to a number of snake species whose venoms are being studied by Hayes and his students. When you step into the room where the snakes are kept, bone-chilling rattles similar to the one I experienced behind Chip's house surround you from all sides. I watched as Chip went about the daily tasks of maintaining the animals, including cleaning the enclosures they are housed in. To do that, Chip had to remove a snake using snake hooks and place it safely in a large trash can while he removed the animal's waste products and cleaned water dishes. I was simply awestruck at the confidence he displayed handling such large snakes whose bites could kill or disfigure him so easily.

Then Chip's labmate David Nelsen took me to another room,

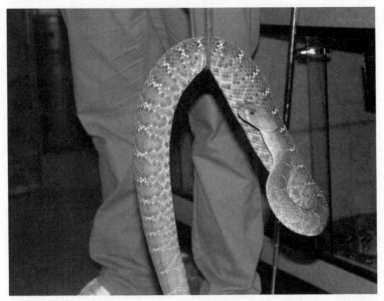

Chip Cochran moving a large red rattlesnake out
of its enclosure at Loma Linda University
(Photograph by Christie Wilcox)

where the researchers house the other venomous animals they
keep in various sizes of Tupperware containers with small holes for
air. The lab doesn't focus just on snakes; they also study scorpions
and spiders. David pulled one container off the shelf to show me
its inhabitant: a large black widow. There must have been one
hundred spider containers covering the shelves from floor to
ceiling. I felt my stomach tighten—spiders, too, are known for the
necrotizing qualities of their venom.

Every year, millions of people show up at their doctors' offices
with what they assume are spider bites that have opened up into
large, leaking wounds. Most don't actually know their attacker.
(If you study these things, it's not hard to guess which group of

spiders is responsible if a verified spider bite festers into a gaping sore. The recluse spiders, including the brown recluse, *Loxosceles reclusa*, are known for two things: their shy nature—hence the name—and their potent necrotic venom. The lesions and other symptoms that result from their bites are medically referred to as loxoscelism—don't Google it . . . *trust me.*)

The wound from a recluse spider starts innocently enough: just a small pair of holes where the spider's venom-laden chelicerae (or mouthparts) pierced the skin. The capillaries around the site begin to constrict and slow blood flow, then fall apart. Within three hours, white blood cells have migrated to the wound, infiltrating the tissues where the venom was injected. The skin swells and becomes itchy and inflamed. In the center of the forming lesion the skin appears blue surrounded by a ring of white (from a lack of blood), then red—a bull's-eye of tissue death. Death *hurts*. Slowly, areas that were once red turn purple, then black, as the flesh dies. In some cases the dead area forms a hard ulcer that later falls off, exposing raw flesh. The process is described in the medical literature as liquefication—"*liquefactive necrosis.*"

Though recluse bite wounds can become large and disfiguring, they usually heal on their own with time. Occasionally, skin grafts are necessary. Loxoscelism can also be systemic; in up to 16 percent of cases, the gangrenous wound can be accompanied by fever, nausea, vomiting, weakness, anemia, coma, and, only in the rarest of cases, death.

Western medicine was completely unaware of these spiders' potent bites until the late nineteenth century, when the first necrotic lesions were described from bites in Tennessee and Kansas. By the mid-twentieth century, we knew that members of the genus *Loxosceles* were to blame, and soon, reports of the terrible wounds that could result from their bites became commonplace. But even now, there is no consensus on the best treatment for loxoscelism. What we do know is that the tissue loss is due largely to a single component of their venom, an enzyme called sphingomyelinase D. Removing this protein from the venom reduces the dermonecrotic activity by 90 to 97 percent. Sphingomyelinase D's main action is to chop up sphingomyelin, a lipid commonly found in your cell membranes. The exact pathways that follow upon this lipid cleavage aren't well understood, but the end result is clear: massive activation of the immune system.

We like to think of our immune system as our body's guardsmen protecting us from trespassers and unsavory folk who have no business being there. Unfortunately, our various immune cells are more like mercenaries: they willingly fight our battles, but given the right incentive, they can easily switch sides. Sphingomyelinase D essentially hands the immune system a briefcase of cash and instructs it to fire at will, and it does—killing the very tissues it once protected.

Sphingomyelinase D activity is found in the venoms of a few other spiders, including the six-eyed sand spider. Recluses and sand spiders are the only two genera in the family Sicariidae, so it's not surprising they share similar venom components. But other than the Sicariids . . . that's it. No other spiders—no other venomous animals of any kind, for that matter—have this particularly potent necrotic enzyme. And it's not just lacking in venoms— sphingomyelinase D activity is completely unknown *in the entire kingdom of animals* except for this one spider subgroup. It is, however, found in pathogenic bacterial species. Some scientists were so struck by the presence of a bacterial toxin in these spiders that they wondered if the spiders managed to steal bacterial genes and make them their own, a process known as horizontal gene transfer. But recent genetic analyses determined that the spiders developed their potent necrotic enzyme independently.

It turns out that while we immediately associate skin lesions with spider bites, rarely are our open sores the work of tenacious arachnids. Very few spiders possess mouthparts strong enough to penetrate the human skin and envenomate at all. And those that do rarely have the chemical weaponry to cause that kind of damage. Some of the most dangerous spiders in the world, including the black widows of America and the redbacks of Australia, often produce only minor effects at the site of envenomation. And the spiders don't like to bite: David Nelsen performed experiments with his roomful of spiders, and found that black widows bite and envenomate only when they are pinched or squeezed, or truly fear their own death. Even "the deadliest spider in the world," the Sydney funnel web spider, isn't known for its local tissue damage. Like that of the redbacks, its venom contains potent *neuro*toxins used to paralyze prey. Despite what you might think, the recluses and their relatives are the only group of spiders known to inflict a necrotic bite with any regularity.

When patients say they have a spider bite (whether they actually saw a spider bite them or not) and there's a skin lesion, the brown recluse is often fingered by victims and doctors alike without hesitation. Studies of bona fide recluse bites paint a very different picture of their overall danger. Only about a third of bites result in the skin necrosis, or necrotic arachnidism, that these spiders are known for. Instead, the vast majority of bites are unremarkable and heal readily on their own. Scientists aren't sure why some bites fester into large, rotting lesions, but likely the health of the victim, location of the bite, and the size and even sex of the spider (female *Loxosceles* venom is almost twice as potent!) all factor in. And let's be clear: they're called *recluses* for a reason. One family in Kansas managed to live with more than two thousand brown recluses for years without a single bite until they finally decided to rid the house of the potential threat.

Most of the so-called spider bites that we bring to the attention of medical professionals are nothing of the sort. In one study, less than 4 percent of supposed spider bites were in fact legitimate; more than 85 percent were bacterial infections. In another, 30 percent of people who thought they had spider bites actually had the potentially deadly methicillin-resistant *Staphylococcus aureus*, also known as MRSA. According to doctors, people have mistakenly blamed spiders for necrosis caused by anything from herpes and syphilis to fungal infections, Lyme disease, cowpox, and even anthrax! Since bacteria possess similar skin-eating enzymes, it's not surprising that we often mistake their presence for bites from the most notorious spiders. Such misdiagnoses can be costly or even deadly: while there is no good therapy for venom-induced necrosis, bacterial infections may be readily treatable, and those that aren't easily treated need to be identified to prevent both the patient's untimely demise and spread of the resistant strain.

Rattlesnakes, too, have a reputation to fight against. It wasn't always that way—back in the late 1700s, the serpents were lauded as a symbol of the American spirit. "Don't Tread on Me," the Gadsden flag reads under the image of a coiled rattlesnake. Exactly when and who first promoted the rattlesnake as an American symbol is unknown. In 1752, Benjamin Franklin quipped that perhaps the most appropriate way to thank the British for sending felons to America was to send rattlesnakes to Britain. The snake was also central to what is referred to as the first political cartoon in an American newspaper: the image of a snake cut into pieces for each of the colonies, with the words "Join, or Die." Over time, the snakes became symbolic of America itself. Rattlesnakes—usually depicted as a coiled timber rattlesnake (*Crotalus horridus*)—began to appear on uniform buttons, paper money, banners, and flags. The first written account of the "Don't Tread on Me" image comes from a letter to *The Pennsylvania Journal* in December 1775, written anonymously (though many scholars believe the author was Ben Franklin). In it, "An American Guesser" remarked on the fitness of the snake as a symbol (which, in this case, had been painted on the drum of one of the first marines):

> I recollected that her eye excelled in brightness, that of any other animal, and that she has no eye-lids. She may therefore be esteemed an emblem of vigilance. She never begins an attack, nor, when once engaged, ever surrenders: She is therefore an emblem of magnanimity and true courage. As if anxious to prevent all pretensions of quarreling with her, the weapons with which nature has furnished her, she conceals in the roof of her mouth, so that, to those who are unacquainted with her, she appears to be a most defenseless animal; and even when those weapons are shown and extended for her defense, they appear weak and contemptible; but their wounds however small, are decisive and fatal. Conscious of this, she never wounds 'till she has generously given notice, even to her enemy, and cautioned him against the danger of treading on her.

Though the marines were among the first to adopt the imagery, they weren't alone. Before the official American flag was chosen, each militia hoisted a flag of its choice, many of which depicted a rattlesnake and "Don't Tread on Me." The First Navy Jack also proudly displayed an uncoiled rattlesnake and the iconic phrase. In 1778, the Continental Congress officially adopted the rattlesnake design as the official Seal of the War Office. The United States Army has officially used the rattlesnake symbolically ever since.

But in the early part of the twentieth century, perceived danger from rattlesnake bites led to the first rattlesnake "roundups," or "rodeos," which have become tourism-boosting annual events in many southern and midwestern states. Hundreds of thousands of rattlesnakes are captured or killed in these events every year. The roundup in Sweetwater, Texas, has been held annually since 1958 and kills upwards of 1 percent of all the rattlesnakes in the entire *state*. American snakebite statistics don't even begin to justify such extraordinary removal efforts. Instead, the tens of thousands of participants put themselves at greater risk of a necrotic bite by engaging so many of these otherwise shy snakes.

We often single out species for what makes them unique. Certainly, venomous animals have their distinctive traits. Who doesn't know a cobra by its hood, or a jellyfish by its tentacles? However, venomous animals and venoms across species, even across phyla, are shockingly similar. Venoms with the same uses often have very similar components, even if the animals that wield them are very different. Hookworms, leeches, snakes, and ticks all possess compounds in their venoms that inhibit platelet aggregation to stop blood clotting. Sometimes such similarities are superficial and accomplished by very different molecules. But often, the same families of proteins have been recruited and revamped many times to serve as venomous toxins by disparate lineages. Tissue-destroying phospholipase A_2's (which chop up membrane lipids) are wielded by cephalopods, cnidarians, insects, scorpions, spiders, and reptiles (and the reptiles recruited phospholipase A_2's *four separate times* in their evolutionary history!). Meanwhile, snails, jellies, corals, worms, insects, scorpions, reptiles, and spiders all employ Kunitz-type peptides, which act as inhibitors of other proteins.

Approximately sixty thousand protein families have been identified to date. Yet the same handfuls of families appear over and over and over again in venoms, from the cnidarians all the way to the primates. That's a lot of coincidence, if we are to believe that the proteins used for venomous purposes are chosen by chance. Instead, the repetitive nature of venom recruitment suggests that some proteins are easily co-opted for nefarious purposes, while others are ill-suited to such physiological dirty work.

What makes a good venom protein? It's an important question for venom scientists to answer. As modern technologies make it easier to detect proteins in venoms, it becomes more of an imperative to determine which cause direct harm and which might not be as pathogenic but instead serve other functions, such as protecting the animal from its own toxins or simply maintaining venom gland cells. Scientists can then focus their medical efforts on combating those that cause the most pathology in us.

There are a few factors that set the venomous proteins apart from the rest. First and foremost, they're *secreted*. In fact, every single venom toxin ever identified is a secretory protein, complete with a specialized signal at one end of the sequence, called the N-terminus, which has to be cleaved off before the protein can act. That's not to say that the protein has to be derived from secreted ancestors—it's possible, and even likely, that venom proteins are sometimes derived from enzymes that once were more tied down to a cellular membrane, but, after gene duplication, have transformed, through either recombination or the relocation of mobile bits of DNA called transposable elements.

In addition to being secreted, all venom toxins perform fundamental biochemical actions. They either (a) chop up molecules found in all living cells, (b) mimic signaling molecules, or (c) compete with body compounds for receptors. The necrotic enzymes, including hyaluronidases, phospholipases, and metalloproteases, are all cutters. They cut apart important things, causing direct and serious damage. Other hemotoxic proteins function as signals or competitors because they are duplicates from the same protein families as the proteins they mimic or block. What better way to inhibit platelet aggregation, for example, than to recruit a platelet inhibitor in the first place?

Most toxins are also fast-acting because they come from short-term physiological processes. You don't see toxins from the families that trigger cell growth or provide structural support, as the process of growing tissues is slow and steady. Instead, venom components need to do their job quickly. If they acted slowly, then the venom wouldn't work. A slow predatory venom means the prey escapes. A slow defensive venom means its wielder becomes dinner. The need for speed, universal applicability, and secretion make immediate sense when it comes to a venom toxin.

Less obvious are some of the other commonalities among venom toxins, like the fact that most are biochemically stabilized. There are many ways in which proteins fold and can maintain their proper folding pattern, but toxins seem to be especially partial to one: disulfide cross-linking. Such atomic bridges are created using cysteines, one of the twenty essential amino acids. Cysteines are common in many secreted proteins because they make molecules less likely to degrade or be cut apart by enzymes, but there are lots of secreted proteins—like the globular enzymes—that use other means of staying in one piece. Yet the toxins strongly favor the cysteine path over others, suggesting that disulfide cross-linking is a biochemically vital aspect of venom proteins.

Venom toxins also tend to come in bunches. Once one venom toxin gene is recruited, it's duplicated again and again, each new gene modified slightly, sometimes taking on completely new activities. A single species may have hundreds of copies of the main toxin genes.

There are some exceptions to the latter "rules," of course, but overall, they hold. These simple commonalities suggest that there are fairly strict biological and biochemical constraints to what can and what cannot act in a venom. It also means that when it comes to managing venomous activities, there are a limited number of targets that pharmaceuticals or therapies need to lock onto. It's a prospect that the next wave of medical scientists is very excited by. If they can identify and focus on the most damaging venom components and create targeted therapies, scientists might not just be able to treat the most dangerous bites; they might be able to create one antivenom that treats them all using antivenomics.

Scientists are particularly keen to make a universal antivenom for the compounds I'll talk about next. While rattlesnakes warn of their presence with that unforgettable sound, snakes whose venoms don't produce such gruesome wounds usually give no such warning before they strike. The lack of damage at the site of the bite itself lulls victims into a false feeling that their injury is mild, when, in reality, far more deadly toxins have been introduced. Hemotoxins are gruesome, but neurotoxins kill quickly, silently, and frequently.

7

DON'T MOVE

No bite appears on the flesh, no deadly swelling with inflammation, but the man dies without pain, and a slumberous lethargy brings life's end.

—NICANDER

Painless. That's the word used by victims to describe the bite of a small octopus found in Australian tide pools. Only about the size of a golf ball, with patterned brownish to yellowish skin, the species in the genus *Hapalochlaena* are all shy, preferring to avoid contact with humans, or pretty much anything else larger than they are. They spend their days hiding, using their color-changing skin cells (chromatophores) to blend in with their environment, or tucking their gelatinous bodies into crevices in the rocks and reef. Though the species are modest in size and temperament, they are armed with some of the most potent venoms in the world. All of them are referred to by the same common name, which comes from the distinctive skin pattern they display when afraid: deep, peacock-blue circles. Blue rings of death, as they have been called. A final warning from the blue-ringed octopus.

Some of their larger eight-legged relatives are known for terribly painful bites, but not the blue-ringed octopus. The two tiny holes caused by their small, parrot-like beaks feel at most like a needle prick or a pinch. Some who have been bitten know it only because of tiny dribbles of blood from where the sharp, chitinous mouthparts quickly pierced. Painless. But deadly.

Anthony and his twin brother didn't know it was dangerous—they were, after all, only four years old in 2006 when they discovered a small octopus in the rocky tidepools at Suttons Beach in Queensland, Australia. The twins' mother, Jane, said she saw Anthony playing with and holding the little animal shortly before he complained that it had bitten him. The young boy began vomiting almost immediately. His vision blurred and he quickly lost the strength to stand. "He said to me, 'I can't walk,' and his legs were all floppy," she reported. Luckily, he was quickly in the hands of emergency responders who knew exactly what was causing the sudden and severe symptoms (though bites aren't incredibly common, the blue-ringed octopus's reputation is now well-known to first responders in the areas where the deadly cephalopods are found). They rushed him to the hospital as he labored to breathe. He soon lost control of his muscles altogether, and had to be transferred to pediatric intensive care. Less than thirty minutes after the bite, the boy was dependent upon a ventilator for survival. It would be another fifteen hours before Anthony's body managed to clear the toxin enough that he could begin to move his muscles on his own, and more than a day before he was strong enough to be discharged.

If he had been any farther from medical care, or could not describe his assailant, Anthony likely would not have survived. A similar creature quickly killed a full-grown man just more than half a century earlier, before first responders knew that blue-ringed octopuses were armed with such potent venoms. Kirke Dyson-

Holland—or "Dutchy," as he was known to friends—was twenty-one in 1954 when he spent the day spearfishing with his friend John about three miles from Darwin, Australia. They were strolling along the tide pools near the beach when they spotted a little reef octopus. Believing it to be good bait for future fishing efforts, John picked up the animal and let it crawl over his arms and shoulders. He and Dutchy had played with octopuses before and never had cause for concern. John soon handed the small creature to Dutchy, who also let it wander around him as he walked until it made its way to the back of his neck. Dutchy didn't feel the brief bite, but a few minutes later, his mouth went dry. As he walked away from the water's edge, he started vomiting, struggled to breathe, and fell to the sand. John got him off the beach and to the hospital. One of the last things Dutchy said to his friend before losing the ability to speak was "It was the little octopus, it was the little octopus." He died two hours later.

At the time, Dutchy's death was presumed to be from an allergy to the octopus's saliva—an unfortunate medical complication. Even a decade later, when Bruce Halstead compiled and wrote one of the most-cited tomes on toxic animals ever published, *Poisonous and Venomous Marine Animals of the World*, little was known about the venom of octopuses and their relatives, and life-threatening bites and deaths were considered aberrations. But in 1970, the Australian scientists Shirley Freeman and R. J. Turner isolated the most lethal component from the venom of one such little octopus (*Hapalochlaena maculosa*), calling it maculotoxin in the absence of knowledge of its chemical composition. When injected into rats and rabbits, the compound caused steep drops in blood pressure and heart rate, and could completely paralyze the animals' respiratory systems. Eight years later, scientists confirmed that maculotoxin was, in fact, the same compound as is found in the flesh of pufferfishes: the infamous tetrodotoxin.

Tetrodotoxin is among the deadliest compounds known to man. It's more potent than arsenic, cyanide, or even anthrax. It's 120,000 times as deadly as cocaine and 40,000 times as deadly as methamphetamine. Like many of the world's most lethal compounds, it is a neurotoxin—it targets our body's nervous system. Unlike the hemotoxins in rattlesnakes and spiders, neurotoxins are fast-acting killers because they numb and incapacitate by inhibiting communication between cells.

The cells in our bodies have several ways of communicating, but the fastest is through electrical signals. Electricity isn't just in wires and batteries; it is, by definition, the energy that results from the existence of charged particles. As kids, we learn that all things in our universe are made of atoms, which in turn are made from three particles: protons (which carry a positive charge), neutrons (which have no charge), and electrons (which carry a negative charge). We can assign a +1 for every proton and a −1 for every electron, and when we sum up all the positive and negative charges in a given atom, we can determine whether it is "charged," either positively or negatively, by whether the total is greater or less than zero. The same is true for molecules, and we refer to these charged atoms and molecules as ions.

Ions want to interact with one another; positives are attracted to negatives, negatives are attracted to positives, and both repel their own kind. So any time you have a barrier where there is more charge on one side than the other, there is the potential for movement should that barrier break down. That's literally what voltage is a measurement of—*potential* energy caused by differences in charge. This is how batteries work: in a 9-volt battery, there are nine volts worth of difference in charge from one chamber to the other, and you can use that potential energy when you connect the chambers with a wire. The electrons in one chamber—the anode—are just busting to get away from one another and find a positive buddy, so the minute you let them move by connecting the battery to a circuit, they do. But nothing happens if you take two batteries and connect negative to negative, because there is no difference in potential across them.

Your cells are essentially microbatteries, with the cell membrane acting as the barrier between two differently charged solutions. Inside your cells, there are more potassium ions (K^+), while outside, there are more sodium (Na^+) and chloride (Cl^-) ions. There are others, too, but these three are the biggest drivers of the difference in charge across cellular membranes. Your average cell has a resting potential of -70 millivolts (mV), which simply means that there are slightly more negative charges inside a cell than outside. Cell membranes maintain this potential constantly, using energy to actively pump out sodium ions that sneak in, and pump in potassium ions that leak out. More important, the nervous system uses these membrane potentials to send lightning-fast signals to and from different parts of the body by way of long, skinny cells called neurons.

To type this paragraph, my brain has to send signals to the muscles in my fingers to tell them when and how to move. It doesn't take seconds or minutes to do this; our neurons can send signals at a speed of about 150 meters per second, which means it takes about 1/150th of a second for my thoughts to reach my fingertips. And just as quickly, I feel the pressure of the key as I push. That speed is vital for coordinating our large, complex bodies, and it would be impossible if it weren't for cell membrane potentials.

When my skin cells touch the keyboard, the pressure causes mechanoreceptors just beneath my skin's outermost layer, the epidermis, to activate, opening force-sensitive ion channels. The instant a channel opens, ions start moving. Ion channels can be general, allowing any charged particle through, or can be shaped and charged very specifically to allow only a single kind of ion to pass. These force-receptive ones, for example, are sodium channels, thus allowing a sudden influx of sodium into the cell. For a brief moment, there is more positive charge inside than out, and the potential of the small section of membrane around the ion channel that was opened when I touched the key becomes +30 mV. Then the force-gated channel shuts.

Nearby on the membrane are similar ion channels that respond to changes in voltage rather than force, so when sodium ions flood into the cell, they swing open. Some of these are potassium channels that serve to return the membrane to the way it was—positive outside, negative inside—by letting potassium rush out of the cell. When the membrane is back to the way it should be, these shut, and active pumps on the membrane slowly return the sodium and potassium ions to the side of the membrane where they belong. All along the membrane there are these voltage-gated channels, so once one is triggered, the movement of ions sets off the next closest. That voltage-gated sodium channel swings open, and the whole process is repeated just a teeny tiny bit farther down. This domino-like cascade is how the electrical signal moves from one end of a long neuron cell to the other, eventually telling my brain that my finger felt something. While it sounds like a lot of activity, these ions travel at very fast speeds across very small distances, and the opening and closing of ion channels happens very, very, *very* quickly.

Through a diverse combination of receptors on one end of a neuron, we are able to *sense*; our noses and mouths have receptors that let sodium flood in when they bind a particular molecule, hence our experiences of smell and taste. Our ears have receptors that are hypersensitive to the tiniest changes in pressure, allowing us to hear. Our eyes have light-sensitive receptors that pick up particular wavelengths of color. Our skin has a diverse array of receptors that respond to many varieties of pressure, temperature, and vibration. These voltage cascades that travel down neurons—referred to as action potentials—also come from the brain. When we move muscles, neurons in our brains translate our very thoughts into these traveling potentials, eventually telling muscle fibers to clench or release. And at the core of all these, from our senses to the movements of every muscle in our bodies, are the ion channels that perpetuate the cascades of electrical energy. These ion channels are what neurotoxins attack.

Tetrodotoxin, for example, is a sodium channel blocker. When a blue-ringed octopus bites, tetrodotoxin in the venom shuts down the victim's neuronal signaling, leading to numbness radiating from where it entered the body. Nausea, vomiting, and diarrhea follow. Weakness and paralysis aren't far behind; when sodium channels in neurons are prevented from perpetuating action potentials, the brain simply cannot tell the muscles to move. Even breathing requires electrical signaling—tetrodotoxin slows and eventually stops the diaphragm altogether. At high enough doses, the victim's heart is unable to beat.

But tetrodotoxin isn't a universal toxin. That's because not all sodium channels are the same: even within our own bodies there are many variations on the same general theme. Tetrodotoxin can be deadly because it strongly binds to a range of sodium channels that are found crucially throughout our bodies and those of other vertebrates, but there are other variants of sodium channels that it isn't effective against. The blue-ringed octopuses are wholly un-affected by their own venom component, and they're not the only species. Tetrodotoxin is wielded either as a weapon or a defense by salamanders, frogs, crabs, sea stars, and even snakes in addition to the pufferfishes, and all are either resistant or immune to its effects. Their bodies have few or no ion channel types that are affected by tetrodotoxin.

Just how many channel types are there? A lot more than you'd think. Potassium channels, for example, are made from a combination of four protein parts. We have seventy or so different genes for these parts, so if you made a channel using four exact copies from one gene, we'd have seventy different potassium channels. But it turns out that combinations work, too—you can have three of one and one other, or all four pieces completely different. Theoretically, there are upwards of 24 *million* different combinations, though scientists have yet to determine if all of them exist or what they do. What we do know is that our bodies employ a plethora of them, each with a different purpose. Some are found only in brain neurons, while others dominate the motor neurons that talk to muscles. Channels for other ions are similarly diverse.

Venoms from diverse species contain neurotoxins that affect every step of our neuronal signaling. Some, like tetrodotoxin, shut down critical channels, while others pry them open. Some stop signals at the very beginnings or ends of these pathways. Some indiscriminately act on a broad range of a given channel type, while others are incredibly specific. Tetrodotoxin is an example of the former—a brute of a neurotoxin, which is lethal in most animal species because it acts on many different types of sodium channels. On the other end of the spectrum are the toxins employed by another group of marine mollusks—compounds known for their elegance, for each has a very specific molecular target. Like artisanal mixologists, these snails craft intricate, unique venom cocktails that precisely incapacitate their prey. And when it comes to paralysis, no group compares to the impressive cone snails.

Tide pools in Hawaii don't have blue-ringed octopuses to worry about, but they still have plenty of dangerous animals with neurotoxic venoms. Cone snails are fairly common in Hawaii, so

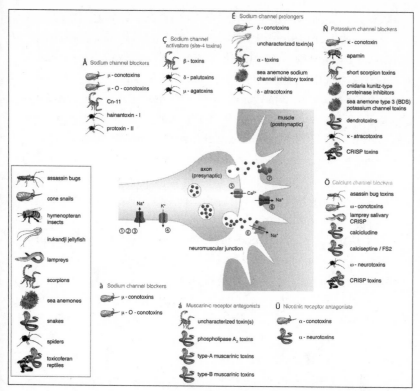

A depiction of a motor synapse and the sundry targets that toxins from
different venomous clades attack
(Figure © Bryan Grieg Fry)

every year, when I work with second graders and their parents in
tide pools, teaching them about marine life, we come across them.
Their shells are distinctively shaped—hence the name. Even when
they are encrusted with algae, it's not hard to tell them apart from
the other snails that inhabit the shallow waters near the beach.
Every year we warn the local kids not to touch the cone snails, for
while it's easy to spot them, it can be difficult to distinguish the
really dangerous fish-hunting species of *Conus* from the just-
gonna-hurt-a-lot worm eaters. And cone snails will often tuck

deep into their mobile homes, tricking you into thinking the pretty shell you just picked up is uninhabited. Yet every year, despite the warnings, I end up wide-eyed and aghast at the potentially deadly species in some kid's hands. I point out to all of them the shape to avoid, *again*, and give my spiel about why they shouldn't be touched, but I wonder if, one of these years, a parent or child will piss off a cone snail enough to find out just how dangerous its sting can be. The snails in those pretty shells can pack potent venoms, ones that still remain a rich source of new toxin sequences for scientists like Baldomero Olivera, even though he's studied them for almost fifty years.

"I had not really intended initially to work on this for very long," he explains to me as we walk through the extensive malacology (mollusk science) collection at the Bishop Museum in Honolulu. The cool, somewhat dark hallways are lined floor-to-ceiling with gray metal cabinets full of shells. Baldomero—or Toto, as he is known by friends and colleagues—is one of the world's experts on cone snails. He first began studying their deadly venoms in the late 1960s. He had graduated from California Institute of Technology with his Ph.D. and spent a couple of years as a postdoctoral scholar, but wanted to return to the Philippines after spending his years of schooling away from home, so he took a position at the University of the Philippines Medical School. Fresh off of working on DNA replication, Toto itched to do high-tech biochemical work, but his new lab in Manila lacked the facilities and funding. As a neuroscientist, he was intrigued by species with toxins that attack neurons. That's when he considered studying cone snails and their deadly venoms. "Because I had collected shells as a kid, I knew that these snails were capable of killing people."

Cone snails are a set of predatory marine mollusks in the family Conidae. They possess a modified "tooth" of sorts shaped like a

The deadliest cone snail in the world—or what remains of one—
in the Bishop Museum's Malacology Collection
(Photograph by Christie Wilcox)

harpoon that is connected by a thin tube to a venom sac. When they attack, they launch this needle-like harpoon into their victim and pump venom through the tube. Their venoms cause near-instantaneous paralysis in their prey. They don't normally strike at humans ("You have to do something fairly stupid to [get] stung," says Toto), but more than a dozen deaths have been caused by their paralytic venoms, most of which are attributed to one species in particular: the geographer's cone, *Conus geographus*—the first cone snail Toto ever studied.

He initially set out to discover what made the geographer's cone so deadly, and began by injecting the venom into the abdomens of mice. The furry test subjects were then placed upside down, clinging to a wire mesh. As the venom spread through their little bodies, it caused sweeping paralysis, and eventually the animals would fall. That was the experimental test (or bioassay) Toto used to isolate the paralytic compounds: falling mice. He carefully separated the venom components through various size or chemical filters, slowly pinpointing exactly what parts of the venom were responsible for the paralytic activity. He was able to narrow it down to peptides—small proteins less than twenty amino acids long, which have come to be known as conotoxins—and after several years of research, he discovered two potent conotoxins: one that acts much like tetrodotoxin, shutting down sodium channels, and one that is similar to venom toxins found in cobras.

He had solved his quest, and had enough funding to afford better equipment. It was time to move on. And, in fact, in response to martial law imposed by the Marcos regime, Toto relocated his young family to the University of Utah in Salt Lake City in 1973. There, he put his cone snail research on a back burner, tended by students. But then, Toto explained, a nineteen-year-old student named Craig Clark (who would become a neurosurgeon) changed everything. Craig had an out-of-the-box idea: instead of just injecting the venom components into the mouse's body, he thought they should inject them into the mouse's central nervous system. Craig went on to do what he proposed, and made a startling discovery: many of the venom components overlooked in the falling test had effects in mice when injected into their brains, and the effects were *weird*. Some peptides caused mice to jump and twist. Others made them run in circles. One even put them to sleep, but only if they were less than three weeks old—it made adult mice run from one corner of the enclosure to the other.

With fancy new equipment, including an HPLC that allowed for fine-scale separation of each peptide, Toto and Craig discovered an unexpected diversity of conotoxins. "This makes them depressed. This makes them comatose. This causes trembling. This causes scratching," Toto continued, showing me on his laptop each of the peptides studied in his lab. "Suddenly we realized that the venom was not a few paralytic toxins but was this incredibly complex diverse pharmacological mix . . . That experiment changed everything."

After Craig's incredible insight, Toto went on to host a "troop" of undergrads. Each could choose any cone snail they wanted; then they would use Craig's test and purify the conotoxin that caused the activity of their choice. One undergrad in the early 1980s decided to purify a compound he called the shaker peptide, because the mice that received it would have characteristic tremors. His name was J. Michael McIntosh, he studied the magic cone, *Conus magus*, and the peptide he discovered was called omega-conotoxin MVIIA. Nowadays, though, it's known by a different name: Prialt, the first FDA-approved drug from cone snail venom.

Toto's eyes lit up as he began to describe how Prialt works, unable to conceal his excitement about his student's discovery. He showed me a video of a synapse at a neuromuscular junction, where neuron meets muscle to tell it when to move. When the electrical signal reaches the end of the neuron that signals muscles to move, it triggers a different kind of ion channel—a voltage-gated calcium channel—to open. This flood of calcium into the cell leads to the release of the neurotransmitter acetylcholine, which then triggers the chain of events that leads to contraction. Prialt, Toto explained, blocks those calcium channels. No calcium flow = no muscle contraction = paralysis.

"How could this possibly become a drug, right?" he asks, with a knowing smile. It all goes back to the fact that cone snail venom peptides are very, very specific. Our calcium channels that trigger muscle movement are different from those found in fish, so the toxin doesn't affect ours. We do, however, have calcium channels that are very similar to the ones found in fish muscles—they just play a different role in our bodies: instead of regulating movement, they are an important part of our pain circuitry. As it does to the motor neurons of fish, Prialt completely shuts down the calcium channels at the ends of our pain-sensing neurons, blocking the transmission of the pain signal to the spinal cord. No calcium flow, no message passed to the brain, and no pain—a godsend in chronic pain cases (like patients suffering from certain types of cancer) where even the most potent narcotics are ineffective.

Prialt's usefulness as a drug stems from its specificity: if it blocked any other calcium channels than the one variant that it targets, it would have too many side effects to be a useful drug. But for venoms as such, specificity can be a problem. After all, it's possible to evolve resistance to individual toxins, as the mongooses have to cobra neurotoxins. To evade this problem, the cone snails don't rely on a single, potent paralytic toxin; they make lots of toxins that derail each element involved in contraction—ones that close off sodium channels, ones that seal potassium channels, ones that shut down calcium channels, and so on.

But this is still only scratching the surface of cone snail venom diversity. The toxins that shut down motor neurons are what Toto refers to as the motor cabal ("because cabals are secret societies out to overthrow existing authority," he explained). Every toxin in the motor cabal is able to paralyze a fish, but because they all work by shutting down the neurons directly, they are slow-acting. It takes about twenty seconds for motor cabal cono-toxins to move via the circulatory system around the fish's body and reach enough neurons to stop the snail's meal from wriggling away. That's just not fast enough for a snail to reliably catch a fish. We also know they have faster-acting toxins: videos of fish-hunting cone snails show that when they strike, their meal is frozen in the blink of an eye.

A completely different set of toxins causes that initial paralysis: what Toto calls the lightning-strike cabal. Instead of turning off channels, this cabal—dominated by the delta and kappa classes of conotoxins—flings them wide open and keeps them that way. The result is a flood of action potentials radiating from the sting site, as if the fish is getting electrocuted. Muscles throughout the body receive the signal to clench over and over, and the fish becomes instantly rigid throughout.

The lightning-strike–motor cabal combo is just one of the ways fish-hunting snails catch their meals. Other snails have different hunting strategies, like putting whole schools of fish into an insulin coma, and relying on other cabals to precisely manipulate their prey. And recently, scientists discovered that cone snails don't just make toxins to kill—they make a different set of toxins to defend themselves with, and are able to wield these distinct predatory and defensive venoms as circumstances require.

Suddenly, it makes sense why each cone snail species has so many conotoxins. But *how* did they come to have such diversity? No two cone snail species possess the same peptides; each has its own unique set, and there are more than five hundred species in the genus *Conus*, more species than any other genus in the oceans. And the cone snails are just the beginning: Toto estimates that if you look at the extended family tree of the cone snail, there are more than ten thousand venomous marine snail species on the planet, each with anywhere from a few hundred to several thousand different toxins. Most of them, including the incredibly biodiverse turrid snails, have never been examined in the laboratory because they are smaller than an inch long and live in areas that are less accessible than the habitats of the shallow-water-loving cones. But they, too, are chock-full of venom peptides. By that count, there may be anywhere from 300,000 to 30 *million* different toxic peptides just waiting to be discovered and sequenced. That kind of venomous success is unmatched by any other toxic group.

So how do the venomous snails create so many toxins? Their secret is a genetic advantage: their venom genes are among the fastest-evolving DNA sequences on earth. We tend to think of evolution by its outward effects: the diversity of traits and behaviors that set each species apart from its closest relatives. But evolution is a measure of genetic change, not physical differentiation. After all, individuals in a species can look very different from one another, but they remain tethered by their genetic makeup. For about a century, scientists have defined evolution as a change in the frequency of gene variations (called alleles) in a population. The "rate" or "speed" of evolution, then, refers to how fast genes mutate or duplicate. And when it comes to evolutionary speed, the venomous snails are the Usain Bolts of the animal kingdom. And you thought snails were slow!

Genomes are the blueprints for life; each contains the full set of genes required to create an animal. Those genes are written in DNA, which is a four-letter language, as there are four different bases of DNA: A for adenine, T for thymine, G for guanine, and C for cytosine. These four letters are arranged in three-letter words, or triplets, called codons, which spell out different amino acids, the building blocks of proteins. AAA codes for the amino acid lysine, for example, while GAA codes for glutamine. But though there are sixty-four different possible combinations of A, G, C, and T, they encode only twenty amino acids, so some words translate to the same protein building block. Because of this, you can change certain bases without changing the meaning; AAA can be mutated to AAG, but either way, it still spells lysine. Changes that alter the amino acid composition of the gene's protein product are called non-synonymous substitutions, while those that don't are synonymous.

Animals can alter their genetic material in one of three ways: (1) mutations, which alter a single letter in the genetic code; (2) insertions/deletions, or indels, which can add or remove anywhere from one base to long chunks of sequence, and even shift the reading frame of a gene—with what base a string of triplets starts, and therefore what its codons spell—if the number of bases in the indel is not divisible by three; and (3) duplications, in which entire genes can be accidentally cloned. Duplication is considered an essential component of evolution because duplicated genes are redundant—the animal worked fine with just one copy, after all, so the one gene continues to fulfill whatever essential role it had before it gained a twin, which leaves the new, second gene free to mutate and insert or delete bases. If you look at our bodies, we have many genes that were duplicated over and over, each new copy gradually diverging from the last to create a new protein. For example, hemoglobin, the molecule that carries oxygen in our blood cells, and myoglobin, the protein that binds oxygen in our muscles, are the products of gene duplication.

Conotoxin genes are among the fastest-duplicating on earth. The A-superfamily conotoxin genes, for example, spontaneously clone themselves, on average, 1.13 times every million years. That's three times as fast as the fastest-duplicating genes from whole-genome studies of other animals, and it's at least twice as fast as the rates for genes renowned for their evolutionary speed, like the genes which encode our ability to smell. And when scientists looked at just the most recent two million years of cone snail evolution, the rate of duplication was even faster: the

A-superfamily conotoxin genes—of which each species possesses dozens—were duplicating at a rate of almost four copies per million years. More important, cone snails maintain this high duplication rate, never settling in. No other animal species is known to have such constantly duplicating genes.

These genes don't just duplicate—they also change exceedingly fast. The rate of non-synonymous substitutions in conotoxins is estimated at between 1.7 and 4.8 percent per million years. That's five times as many mutations as the highest rates reported for mammals, and three times the highest rates found in fruit flies. And that's just the average rate—right after a duplication event, the non-synonymous substitution rate is at least *23 percent per million years.*

While the cone snails set the bar for speed, many venomous animals have toxin genes that clock in with high evolutionary rates, particularly their neurotoxins. Rapid diversification is the hallmark of many venoms, and this accelerated molecular evolution leads to inconceivable toxin diversity—more toxins than scientists have been able to count, sequence, or study. Diversity allows for variation in venoms at every level; there is variation between toxin types in a single individual, variation based on age or sex, variation within species and between them. And yet, venoms from species as disparate as octopuses and snakes converge upon the same targets. The actual molecules in venom are hypervariable, but the functions of those components are markedly conserved.

The drive for evolutionary speed isn't to come up with new targets to attack—it's to ensure that toxins stay potent over time. Neurotoxins, as you can see in the case of the cone snails, are excellent for capturing prey, as causing rapid paralysis is a great way to slow down your potential meal. But the minute a venomous predator starts using a molecule which shuts down its prey's sodium channels, it creates a strong selective pressure in the prey for sodium channels that don't respond to the toxin. Sometimes, as we saw with the mongoose, all it takes is a few mutations to derail a toxin's activity. Venomous animals must always be prepared to change things up, to throw a curveball and hope that they get a strike—or, as the cone snails do, to throw hundreds of balls at once, making damn sure their prey can't hit them all. "What the snails are doing is the equivalent of combination drug therapy," Toto explained. "They don't just use one drug for a physiological endpoint, they always use multiple components."

While the goal is to stay ahead of changes in the venom targets, diversity and accelerated evolution also allow venomous species the opportunity to change diets. That's what happened in the cone snails, according to the newest studies on venom genes. Evidence suggests that the fish hunters, which are the most deadly to us, were able to switch to fish prey only *after* they'd already evolved defensive conotoxins that target vertebrate ion channels! The shift went something like this: Cone snails were once all worm eaters. Fish, as top of the food chain in the oceans, posed a physical threat to the cones, and fast-evolving venom genes gave the mollusks a defensive weapon that allowed them to survive and fight off these piscine predators. Some of the snails' ever-changing toxins became so potent to channels in fish tissues that the fish died, opportunistically offering the snail a new meal option. At least three different lineages of cone snails made the switch over to speedy prey.

Almost every branch of venomous life has at least some neuro-toxic animal up its sleeve—cone snails and blue-ringed octopuses aren't the only ones. Spider venoms are packed with neurotoxins, but they don't tend to cause mammals much harm. Notable exceptions include the widow spiders, especially the notorious

black and brown widows. Their potent neurotoxins—latrotoxins—
don't so much target ion channels as *make* them: these toxins can
form calcium-permeable pores in membranes, making neuronal
synapses fire uncontrollably. Victims who experience systemic
effects suffer extreme pain, cramping, an increased heart rate, and
spasms that can last for days and even weeks. Scorpions, too, are
neurotoxin masters, though most of their toxins are aimed at
nonmammalian targets. A few scorpions, though, possess neurotoxins capable of incapacitating us, most notably the fat-tail scorpion, whose venom can cause seizures and comas; the aptly named
deathstalker scorpion; and the Indian red scorpion, whose venomous sting has been reported to kill from 8 to 40 percent of
human victims, with the highest fatality rates in children.

But you can't talk about neurotoxins without highlighting the
species whose neurotoxins have instilled fear and fascination in
humankind since the beginning of our tenure on this planet; species whose potent bites may have driven the evolution of our eyes
and minds; species that star in the stories of civilizations past
and present, and that even today are among the most recognizable animals on the planet. I'm talking about the elapid snakes,
led by the unmistakable cobras.

Vipers conquer the flesh, leaving bloodied, mangled bodies in
their wake. Elapids, though, are the opposite; sometimes, their
bites aren't even noticed until a thorough autopsy. Almost all of
the deadliest snakes in the world are found in the family Elapidae,
including the cobras, mambas, kraits, taipans, death adders, sea
snakes, and coral snakes. Like the cone snails, these species use
potent peptides to paralyze their intended prey, but unlike the
snails, spiders, and scorpions, their prey items *are* usually mammals,
so it's no wonder that their toxins are lethal to us. We're simply
the taller, lankier, and less hairy relatives of their usual meals. The
most lethal of these neurotoxins are the alpha-neurotoxins, often

referred to as "three-finger" toxins because their folded shape is a core with three loops. These toxins inhibit the neurotransmitter receptors on muscle cells, causing death by paralysis.

What's most intriguing about the venoms of some elapid snakes is that their neurotoxins, though potent paralytics, are reported to have other effects in our bodies. These snakes don't just mess with our muscles—they do something far more sinister and unsettling: they manipulate our minds.

8

MIND GAMES

My body, it was electric. For the first time in my life
I felt as if I had a real heart and a real body and I knew
that there was this fire in me that could have lit up the
entire universe. No book had ever made me feel that
way. No human being had ever made me feel like that.

—BENJAMIN ALIRE SÁENZ

don't know if cockroaches dream, but I imagine that if they do,
jewel wasps feature prominently in their nightmares. These small,
solitary tropical wasps are of little concern to us humans; after
all, they don't manipulate our minds so that they can serve us up as
willing, living meals to their newborns, as they do to the unsuspecting cockroach. It's the stuff of horror movies, quite literally;
the jewel wasp and similar species inspired the chest-bursting hor-
rors in *Aliens*. The story is simple, if grotesque: the wasp controls
the minds of the cockroaches she feeds to her offspring, taking away
their sense of fear or will to escape their fate. But unlike what we
see on the big screen, it's not some incurable virus that turns a once-
healthy cockroach into a mindless zombie—it's venom. Not just

any venom, either: a specific venom that acts like a drug, targeting the cockroach's brain.

Brains, at their core, are just neurons, whether we're talking human brains or insect brains. As I noted in the last chapter, there are potentially millions of venom compounds that can turn neurons on or off. So it should come as no surprise that some venoms go beyond affecting the peripheral nervous system—the neurons that are connected to our muscles and other cells throughout our bodies—to target the carefully protected *central* nervous system, including our brains. Some leap their way over physiological hurdles, from remote injection locations around the body and past the blood-brain barrier, to enter their victims' minds. Others are directly injected into the brain, as in the case of the jewel wasp and its zombie cockroach host.

Jewel wasps are a beautiful if terrifying example of how neurotoxic venoms can do much more than paralyze. The wasp, which is often just a fraction of the size of her victim, begins her attack from above, swooping down and grabbing the roach with her mouth as she aims her "stinger"—a modified egg-laying body part called an ovipositor—for the middle of the body, the thorax, in between the first pair of legs. The quick jab takes only a few seconds, and venom compounds work fast, paralyzing the cockroach temporarily so the wasp can aim her next sting with more accuracy. With her long stinger, she targets her mind-altering venom into two areas of the ganglia, the insect equivalent of a brain.

The wasp's stinger is so well tuned to its victim that it can sense where it is inside the cockroach's dome to inject venom directly into subsections of its brain. The stinger is capable of feeling around in the roach's head, relying on mechanical and chemical cues to find its way past the ganglionic sheath (the insect's version of a blood-brain barrier) and inject venom exactly where it

The emerald cockroach wasp injecting venom into her victim's brain
(Photograph by Emanuele Biggi)

needs to go. The two areas of the roach brain that she targets are very important to her; scientists have artificially clipped them from cockroaches to see how the wasp reacts, and when they're removed, the wasp tries to find them, taking a long time with her stinger embedded in search of the missing brain regions.

Then the mind control begins. First her victim grooms itself, of all things; as soon as the roach's front legs recover from the transient paralysis induced by the sting to the body, it begins a fastidious grooming routine that takes about half an hour. Scientists have shown that this behavior is specific to the venom, as piercing the head, generally stressing the cockroach, or contact with the wasp without stinging activity does not elicit the same hygienic urge. This sudden need for cleanliness can also be induced by a flood of dopamine in the cockroach's brain, so we think that the dopamine-like compound in the venom may be the cause of this germophobic behavior. Whether the grooming itself is a beneficial feature of the venom or a side effect is debated. Some believe that the behavior ensures a clean fungus- and microbe-free meal for the vulnerable baby wasp, while others think it may merely distract the cockroach for some time as the wasp prepares the cockroach's tomb.

Dopamine is one of those intriguing chemicals found in the brains of a broad spectrum of animal life, from insects all the way to humans, and its effects are vital in all these species. In *our* heads, it's a part of a mental "reward system"; floods of dopamine are triggered by pleasurable things. Because it makes us feel good, dopamine can be wonderful, but it is also linked to addictive behaviors and the "highs" we feel from illicit substances like cocaine. It's impossible for us to know if a cockroach also feels a rush of insect euphoria when its brain floods with dopamine— but I prefer to think it does. It just seems too gruesome for the animal to receive no joy from the terrible end it is about to meet.

While the cockroach cleans, the wasp leaves her victim in search of a suitable location. She needs a dark burrow where she can leave her child and the zombie-roach offering, and it takes a little time to find and prepare the right place. When she returns, about thirty minutes later, the venom's effects have taken over—the cockroach has lost all will to flee. In principle, this state is temporary: if you separate an envenomated roach from its would-be assassin before the larva can hatch and feed and pupate, the zombification wears off within a week. Unfortunately for the envenomated cockroach, that's simply too long. Before its brain has a chance to return to normal, the young wasp has already had its fill and killed its host.

The motor abilities of the roach remain intact, but the insect simply doesn't seem inclined to use them. So the venom doesn't numb the animal's senses—it alters how its brain responds to them. Scientists have even shown that the stimuli that normally elicit evasive action, such as touching the roach's wings or legs, still send signals to the animal's brain; they just don't evoke a

behavioral response. That's because the venom mutes certain neurons so they are less active and responsive, leading to the roach's
sudden lack of fear and willingness to be buried and eaten alive.
This venom activity requires toxins that target GABA-gated
chloride channels.

GABA, or γ-aminobutyric acid, is one of the most important
neurotransmitters in insect—and human—brains. If neuron
activity is a party, then GABA is a wet blanket; it dampens a neuron's ability to be triggered through activation of chloride channels. When chloride channels open, they allow negative chloride
ions to flow. Since these ions like to hang out with positive ions,
if these channels are open when a sodium channel happens to open,
chloride ions can cross the membrane at the same pace as sodium
ions, making it harder for the sodium ions to start the domino
cascade that is neuron signaling. Even though a neuron receives
the "go" command, the action potential is stopped in its tracks.
However, GABA isn't a complete inhibitor—the chloride channels can't wholly keep up with sodium channels, so a strong stimulus can overcome the dampening effect. This dulling system is
what the wasp co-opts to make the cockroach do her bidding.
Her venom is packed with GABA and two other compounds that
also activate the same chloride receptors, β-alanine and taurine.
These also work to prevent the re-uptake of GABA by neurons,
prolonging the effect.

While these venom compounds can cut the brain activity
that would make her prey flee, what they can't do is make their
way to the right parts of the cockroach brain by themselves.
That's why the wasp has to inject them directly into the cockroach's
ganglia. Fortunately for her, in a convenient quirk of nature,
the same venom that zombifies roach brains works like magic
to produce the transient paralysis needed to line up the cranial

injection. GABA, β-alanine, and taurine also temporarily shut down motor neurons, so the wasp needs only one venom to complete two very different tasks.

With her prey calm and quiescent, the wasp can replenish her energy by breaking the roach's antennae and drinking some sweet, nutritious insect blood. Then she leads her victim to its final resting place, using what remains of an antenna as an equestrian uses the reins on a bridle. Once inside her burrow, she attaches one egg to the cockroach's leg, then seals her offspring and the roach in.

As if the mind manipulation wasn't bad enough, the wasp's venom has one final trick. While the roach awaits its inevitable doom, the venom slows down the roach's metabolism to ensure it lives long enough to be devoured still fresh. One way metabolism can be measured is by how much oxygen is used up over time, as all animals (including us) use oxygen in the process of creating energy from food or fat stores. Scientists have found that oxygen consumption by cockroaches that have been stung is much lower than that of their healthy roach friends. They thought this might be due to the reduced movement of the complacent victims, but even when paralysis is induced by using drugs or severing neurons, the stung cockroaches live longer. The key to the prolonged survival seems to be hydration. How exactly the venom can act to keep a roach hydrated isn't known, but it ensures that when the wasp larva hatches from its egg, its meal is ready to eat. And soon enough after that, a new wasp emerges from the burrow, leaving the roach carcass behind.

Jewel wasp venom is only one example of neurotoxic venom taken to the extreme. There are more than 130 species in the same wasp genus, including the newly described *Ampulex dementor*, named for the soul-sucking guards of the magical prison

Azkaban in the Harry Potter series. *Ampulex* belongs to a very large and diverse group of wasps, numbering at least in the hundreds of thousands of species, which are known for some serious mental manipulation. All have a macabre lifecycle: as adults, they feed like other wasps and bees, but as larvae, they must feed off other animals. They're not quite independent, not quite parasites—they're parasite-*ish*, or, as scientists call them, parasitoids.

Cockroaches aren't their only targets; there are parasitoid wasps that lay their eggs in spiders, caterpillars, and ants. The temperate Northern Hemisphere wasp *Agriotypus* will dive underwater to attach her eggs to caddisfly larvae, and can remain submerged for up to fifteen minutes to accomplish her task. The brave *Lasiochalcidia* wasps of Europe and Africa throw themselves into the nightmarish jaws of an ant lion, pry them apart, and insert their eggs into its throat. There are even wasps called hyperparasitoids that parasitize other wasps like themselves, such as *Lysibia* species of Europe and Asia, which will sniff out caterpillars parasitized by fellow parasitoid wasps in the genus *Cotesia* and lay eggs in the freshly pupated wasp larvae. In some cases, multiple wasp species parasitize one another, leading to a Russian doll of parasitic interactions.

And to ensure their safe passage from larva to adulthood, these wasps often gain more than just a meal from their hosts. One of them turns its caterpillar hosts into undead bodyguards that will defend pupating young wasps that just ate through its body. Another species' larva forces its spider host to spin it a deformed but durable web to protect its cocoon just before killing the arachnid.

While the wasps in this unusual family may have perfected the art of mind control, there are other venomous species whose toxins alter mental states. There are even species whose neurotoxic compounds get through our own blood-brain barrier, a feat that no wasp venom can yet achieve. But unlike cockroaches, we *Homo sapiens* have a strange affinity for substances that mess with our minds. While the roaches run from those that would twist their brains, some people are willing to pay upwards of $500 for a dose of venom in order to have a similar experience.

If you hit the party scene in the United States, it's not uncommon to see—and be offered—a wide variety of illegal substances, from a joint to a tab of acid. But even these substance-abusing clubgoers can't claim to be as wild as India's rave revelers. In America, a sly smile and a few Benjamins might buy you an eight ball of cocaine, but in Delhi, a similar amount could get you a taste of cobra venom.

According to some people, it's the best high on the market—and for the price, it would have to be. A pinch of powdered venom to spike your drink (referred to by the street names K-72 or K-76) can cost five to ten times as much as other illicit drugs in India. And according to local officials, the drug is so potent that it can give the consumer "such a high that they don't know where they are or what they are doing." At non-incapacitating doses, the venom is said to increase sensations and ramp up energy, similar to a bump of cocaine. The price tag has made it popular among affluent young Indians in particular.

Because of the reputation of its high-class high, smugglers can make a small fortune off of liquid venom from snakes in India, with one liter fetching as much as 20 million rupees—more than $300,000—though about two hundred snakes die in the extraction process. The black-market trade has become such an issue that drug enforcement agencies have started teaming up with wildlife experts to crack down on the illegal sales. Over the past several

years, venom busts have made headlines, featuring criminals caught
with condoms full of snake venom and large glass containers full of
the precious liquid, worth more than $15,000,000. And authori-
ties are now using modern molecular technologies to determine
which species of snakes were milked to create these illicit drugs,
prosecuting smugglers both for drug-related offenses and for violating protected-species laws.

Of course, not everyone who wants a taste of toxin intoxication can afford the stiff price tag of powdered and prepared cobra
venom. Less-well-off thrill seekers are known to obtain their
venom via a more direct route. There are dealers in some cities in
India that sell snakebites for recreational purposes. These dealers
are either stand-alone salesmen or a part of seedy recreational
establishments referred to as "snake dens"—a linguistic homage
to the opium dens of yore—where a person can spend hours in a
bite-induced stupor. The handful of accounts of these dens that
exist suggest they are tucked in dangerous neighborhoods of major cities across India. Some so-called snake dens boast a variety
of snakes to choose from, each graded as providing mild, moderate, or severe effects. The few willing to describe their experiences
say they were offered bites from cobras, kraits, and other elapids.
But regardless of the species involved, the simple fact is that snake
venoms used recreationally, in dens or otherwise, are potentially
deadly, and some are notorious killers with annual death tolls in
the thousands.

Take the case of "Mr. PKD," a fifty-two-year-old man with
more than three decades of drug use, who turned to snake venom
"to experience the kick the other substances now lacked." Rather
than purchasing the pricey powder, Mr. PKD sought the aid of a
nomadic snake charmer who, for the right price, coaxed a snake
to bite him twice in the forearm over a two-week period. The
"high" reportedly began with dizziness and blurred vision, fol-

lowed by "heightened arousal and a state of well-being lasting a few hours"—a better high than Mr. PKD had experienced with opiates. Other cases describe similar feelings: one regular user sought out a snakebite dealer in an urban slum, letting an Indian cobra (*Naja naja*) bite him on the foot to experience "a blackout associated with a sense of well-being, lethargy, and sleepiness." Another long-term substance abuser claimed to take a bite to the toe two to three times a week—he would have indulged daily, but the cost was too steep. And yet another patient described experiencing "grandiosity, a sense of well-being and happiness after each bite."

In 2014, a nineteen-year-old arrested in Kerala, India, admitted to regularly traveling almost one hundred miles to get his serpentine fix. He paid up to $40 to have the small snake's head pressed under his tongue until it bit him, giving him a high that he said lasted for several days. For some, snake venom is their one and only drug. Take the case of two young software engineers with no history of drug use who turned to venom to help calm nerves and fight insomnia. They, too, described bites to the mouth as well as fingers and toes. It's said that bites to the tongue are preferred because they provide a "quick effect" and an "extra kick." One of the engineers even claimed the benefits of snake venom went beyond improvements to mood and a good night's rest: he felt his "sexual desire" also increased.

You would think that a bite would hurt too much to be worth the rush. Some definitely do—viperid snakebite survivors often describe deep, burning pain following the bite and tenderness and swelling that can last weeks. But there are people who have experienced snakebites so mild that they sleep right through them. Some cobra victims in particular question whether they've even been envenomated, as snakes can "dry bite" (in other words, bite without injecting venom), and the onset of symptoms can be delayed for more than half an hour. The surprising consistency among all the accounts of recreational snake venom intoxication of no swelling at the site of the bite suggests that the snakes being used possess venoms with basically no hemotoxicity and diverse neurotoxins—a hallmark of the elapids. Some of the very same species encouraged to bite a tongue or a toe are responsible for thousands of deaths every year. The two engineers even stated that they knew of at least six deaths occurring in the den they visited near Salem, Tamil Nadu, India. With so much at stake, you would think the risk would outweigh the reward—but that would be to underestimate just how amazing the neurotoxins in snake venom can feel.

The high from snake venom is not an immediate effect. Much like taking a double shot of whiskey, once you get it down, it takes a little while before you really start to feel the intoxication. But somewhere between thirty minutes and two hours after a bite, it hits you. The neurotoxins take hold. Dizziness. Blurred vision. And then—euphoria.

In his memoir, Bryan Fry describes the effects of a bite from the aptly named Pilbara death adder as "the most delicious sensation." Though his entire body was paralyzed—artificial respiration kept him alive, as his own diaphragm was crippled by the neurotoxic venom—he "didn't care." As he explains:

> The lights became very bright and the colours very vivid in a way quite like being on psychedelic mushrooms . . . The neurotoxins were now having an extremely potent narcotic effect. Life was beautiful. It was like breathing the most potent dental gas, times a thousand. Once I lost my ability to move at all and

was put on artificial respiration, the sensation kicked up another gear and I was floating high above the world without a single care. True, I was locked inside my body, completely cut off from the outside world—the most primordial of fears. Strangely, I did not mind. This was entirely to do with the fact that I was having the most amazing party-for-one inside my immobile shell. Time warped. For aeons I drifted contentedly through the universe, exploring far-off lands and distant galaxies. This was a classic dissociative out-of-body-experience; a psychedelic state of mind that is reached by disconnecting the mind from the body, either by dissociative drugs like ketamine or, as it turns out, the neurotoxicity of certain toxins. Unlike a bad mushroom trip, however, I did not wonder if it would ever end.

Bryan is lucky to have survived—and certainly wouldn't have without the extensive medical intervention he received. But others claim it doesn't take a near-death experience to feel a sort of "high" from venom. Some self-immunizers, including Steve Ludwin, describe similar (albeit milder) effects from their routine injections.

As I mentioned in chapter 3, the self-immunizers, or SIers, inject themselves with dilute venom to induce the human body's natural immune response. By slowly increasing the venom dose, they hope to stimulate enough circulating antibodies to be effectively immune to the venom or venoms they use, and thus the snakes they keep. Many of them take SI'ing very seriously, including taking notes before, during, and after every injection, describing in detail everything that happens, both physically and emotionally. And though all will say they inject for immunity, not the rush, several SIers say they have had high-like responses to venoms.

While Steve Ludwin doesn't inject for the purpose of getting high, he has said that he feels rejuvenated by very low doses of cobra venom. As he explained, "Cobra venom makes you feel kind of charged." He also compared the sensation to cocaine use: "It's different, but similar at the same time; it seems to make you feel more keen to be active. And physically more able. But it's different because there's not so much of a mood change and no drippy nose the next day." Similarly, Anson Castelvecchi, a self-immunizer and registered nurse whose interest in venoms and other natural therapeutics that have the potential to stimulate the immune system stems from his own struggle with autoimmune ailments, described feeling a "good, clean high" after his first self-injection of copperhead venom. Though the injection (the tiniest amount of milked venom diluted 1:10 in saline placed just under the skin) burned at first, an hour later he felt incredible. He was filled with "tremendous energy," he says, also drawing a comparison to the feeling of cocaine, but "cleaner." He even questions the use of the term *high*, as the effects he felt were "the opposite of impaired."

While Bryan has felt the narcotic effect of neurotoxins first-hand, he is skeptical of reports of recreational snake venom use, especially the idea of a powdered club drug. He says it's unlikely that powdered venom could have any intoxicant effects. Human stomachs can digest proteins with ease—even ones that in our blood would wreak havoc—so dried venom added to an alcoholic beverage should be destroyed by our digestive tract. Indeed, the power of the stomach to tear apart venom has been known for centuries. During the Roman civil war, Cato the Younger is said to have reassured his troops, who were refusing to drink from a spring surrounded by snakes for fear the venom had poisoned it, that "their venom is in their bite, and they threaten death with their fangs. There is no death in the cup." Bryan thinks it's far more likely that the recreational "powdered venom" contains other illicit drugs responsible for the activities reported, and that peddlers are merely calling their product "cobra venom" to make it sound sexier to their clientele.

But even if the powdered version is bogus, Brian Hanley, the founder of and chief scientist for Butterfly Sciences, a private company that works with other self-immunizers to develop vaccines for snake venom, is fascinated by the reports of getting high on actual bites. Based on the description of the effects, he thinks it's possible that the venoms contain neurotoxins that are able to act on brain dopamine neurons, similar to the party drug γ-hydroxybutyric acid (or GHB). He also wonders if the snakes used are a captive-bred line, selecting specifically for no tissue damage and milder venoms. "Snake handlers in India have had thousands of years to do that," he notes.

Back in the United States, Jim Harrison has had a lot of time to think about venoms and their effects on the mind. A well-known and respected member of the herpetology community, Jim is the director of the Kentucky Reptile Zoo, which produces snake venom for antivenom production and scientific research. He has more experience than most with snake venoms; he underwent his ninth accidental envenomation in early 2015 when a snake-restraining tube broke while he was milking a South American rattlesnake, allowing it to free its head and bite his hand. While that might seem like a lot of bites for one man, Jim has milked venom from countless snakes over the past four decades of running the zoo with his wife, Kristen Wiley (he averages about six hundred to one thousand snake milkings a week!). The zoo houses two thousand snakes, most of which are off-exhibit and kept for the express purpose of meeting the high demand for venom to be used in antivenom production.

According to Jim, whose experiences include a bite from an

Indian cobra, the so-called "high" from snakebite, as described by the SIers as well as the Indian medical literature, "is an elapid experience." He's felt firsthand the increase in sensitivity and sensation in response to venom: "You'll feel like everything around the room is really lit up," he explains; "you're aware of everything—people talking, everything else."

While he wouldn't recommend his life-threatening experiences to anyone, his work with opiate addicts in college prior to his current job has given him more insight than most into how addicts think, and he can see the appeal of venom. "Somebody who is an opiate addict is already shooting stuff into themselves that's going to kill them eventually," he says, so the danger isn't a factor. "They want to get that high again, so they'll take a bite or they'll shoot up.

"One of the reasons they seek out opiates in the first place is to get rid of the pain," explains Jim. And there's good evidence that cobra venom contains strong painkillers; experiments in the early twentieth century found low doses of cobra venom could provide complete analgesia in patients with terrible pain that other opiates couldn't alleviate. Though a cobra-based drug for humans has yet to make it to market, you can buy a toxin from cobra venom called cobroxin that is used by veterinarians as a horse analgesic. "They use it to dope up horses and run 'em," Jim explains. It's illegal to use in races, but it's not illegal as a therapeutic. So if the venomous bites are providing pain relief plus a dopamine rush, well, that would be pretty tempting for many of the people who turn to other illegal substances. Since there has been no research directly connecting venom neurotoxins to the "high" associated with bites, some would argue that venom toxins have little to do with the effects Jim and others describe. But perhaps a better explanation is that the chemical cocktails *do* contain neurotoxins able to do something terrifyingly wonderful: cross the blood-brain barrier and affect our minds.

There are several components from venoms known to do exactly that. One of the most researched is apamin, a component of bee venom, which shuts down calcium-dependent potassium channels to make neurons easier to trigger. In high doses, it can cause tremors and convulsions, but at lower doses, something interesting happens. Experiments conducted on rats have shown that apamin injection improves learning and cognitive performance, evidence that it affects mammalian minds. But most telling are studies that show how apamin acts on receptors in dopamine-releasing neurons in the forebrain, making them more sensitive to stimulus or the lack of inhibitory signals. In other words, even when injected into the body, apamin can travel past the blood-brain barrier and prime the brain's reward system—a common feature of recreational drugs.

Snake venoms don't contain apamin (at least we don't think they do). Most of their neurotoxins are much larger proteins and peptides that aren't able to pass into the brain with as much ease— though the key word here is *most*. There is a growing body of evidence that at least some snake venoms contain molecules that can find their way across the blood-brain barrier, and their effects may explain why people feel a rush from bites.

Two hours after injecting South American rattlesnake (*Crotalus durissus terrificus*) venom into the bodies of mice, for example, scientists can detect it in the mouse brains. At least one compound from the venom is known to increase blood-brain barrier permeability, and so it may allow toxins that are normally kept out to make their way into the central nervous system. Another pair of isolated toxins from the same species, crotoxin and crotamine, have analgesic properties that stem from central nervous

system activities, suggesting they either cross the barrier themselves or manage to induce effects from the outside. Several alpha-
neurotoxins from cobra species have similarly demonstrated
painkilling effects that, as far as scientists can tell, occur from direct
action on the central nervous system. The more scientists look,
the more venom toxins they find, from snakes and other species
like scorpions, which make their way past the brain's defenses.

The simple truth is that we don't know all the components of
most venoms. Though studies are discovering more every day,
most still focus on one type of molecule at a time—proteins,
peptides, lipids, or small molecules—because it requires lots of
expensive equipment and specialized knowledge to separate and
sequence each molecule type. And in many cases, scientists are
not really looking to uncover all the venom's components—just
the ones that cause the worst symptoms, so they know how to
treat a bite. "There's got to be stuff in the venom that we don't
understand, and until somebody comes up with a test for it, we're
not going to know it even exists," Jim explains. It's not like the
most recently discovered compounds evolved in the past fifty
years; scientists just didn't know how or where to look for them
until now.

In Suzanne Collins's *The Hunger Games*, one particularly perni-
cious pest plagues the combatants in the battle arena. Called
tracker jackers, these wasps armed with hallucinogenic venom
drive the contestants mad and, later in the series, are used as a
brainwashing tool to alter a character's feelings toward the lead
heroine. In the books, such beasts arose through genetic editing,
but there's no reason to think they couldn't evolve. Indeed, if you
are swayed by Michael Pollan's argument regarding *Cannabis* species in *The Botany of Desire*—that acting as a human hallucinogen

is actually an evolutionary boost, since our drug-seeking species cultivates and concentrates anything that can get us high—then there's reason to believe that tracker jackers may eventually evolve through artificial selection. Indeed, there may be hymenopteran species—those with toxins like apamin, for example—that have already started down a path toward human mind manipulation, and await us in our distant future. For now, though, the wasps are flooding insect brains with dopamine, until the day we harness their potential to do the same for us.

Of course, the only idea scarier than a mind-altering wasp is imagining our own species with such power: Would you trust any government with a toxin that can enhance pleasure, let alone create it? But to blindly fear the unknown neurotoxic contents of venoms discounts their immense potential. Venom toxins could be used as pesticides that kill one and only one species of bug by specifically targeting ion channel variants found only in that species. They could provide effective pain relief that doesn't create dependence and addiction by turning off the right neurons without tripping others. They hold promise as the basis for drugs to not just treat but to *cure* the incurables, including neurodegenerative disorders and cancers. Specificity is gold when it comes to pharmaceuticals, and the incredibly specific targeting of these venom toxins may just allow us to manipulate our bodies in ways we have only dreamed of—and that's exactly what the scientists in the next chapter are counting on.

For thousands of years, we have vilified and deified venomous species, terrified yet fascinated by the toxic cocktails they possess, and only now are we starting to realize how much these potent species have to offer. Millions of years of evolution have honed and fine-tuned venom toxins into highly specific drugs, which can be used to our advantage rather than as weapons against us. We're finally able to turn the tables on some of the deadliest species on this earth by dissecting their cocktails and discovering the lifesaving compounds hidden in their killer bites and stings. What we've learned so far is only a fraction of the biochemical knowledge that venomous species possess—and that little bit, already, is changing *everything*.

9

LETHAL LIFESAVERS

Let us learn from the lips of death the lessons of life.

—FELIX ADLER

Death is what first drew us to venomous animals. We simply couldn't help ourselves; the idea that a limbless snake or a fragile spider could overpower a strong, intelligent primate was so absurd, and yet so terrifying, that our curious species was compelled to look more closely. Such deadly power demanded respect and reverence, and we gave the animals that brandished the most lethal venoms the honors they deserved. But now it's their ability to grant life that will forever entwine our fates with those of the venomous.

When it comes to death, the venomous animals pale in comparison with the world's true mass murderers. Cardiovascular diseases, diabetes, cancer, and chronic respiratory diseases kill more people annually than all other deaths combined. The incredible burden of disease on our health isn't just a consequence of age, either; diseases also dominate deaths under age fifty, and these include notorious incurables like HIV and certain cancers. While we may wish for a good antivenom if we happen to be bitten by a venomous snake, to truly reduce premature deaths worldwide, we need solutions that tackle the killer ailments. And, counterintuitive as it may seem, scientists have come to realize that those medical marvels are awaiting discovery in the deadliest of venoms, because the molecular mechanisms venoms derail are the very same ones we need to manipulate to cure these diseases. By controlling the dose, scientists can turn noxious toxins into miraculous treatments—as John Eng did when he discovered Byetta in the venom of a notorious lizard.

Eng had never seen a Gila monster when he first ordered a sample of its venom in the early 1990s. Most people might prefer it that way, for the nearly two-foot-long lizards have a fearsome reputation in their native deserts in the southwestern United States. It's right there in the name: not Gila *lizard*—an apt moniker for a species of reptile that was once plentiful in the Gila river basin, where it was first discovered—but Gila *monster*. And as with all monsters, there are tales of the Gila that arouse fear in even the bravest of men.

"The writer has never seen a victim bitten, but has heard numerous and gruesome stories of Gila monster bites, and has seen the monster, and would no more readily trust its fangs than he would those of the diamond-back rattler, or the dreaded West India fer-de-lance itself," begins an article titled "The Horrible Gila Monster" in *The Salt Lake Tribune* from 1898. The piece goes on to describe how it is believed that the lizard "can poison its victims to the death by merely breathing upon them as effectively as by biting," and that "its skin exudes a mortal poison." The author is skeptical of those claims, but, he states, "that the bite of the Gila monster has sometimes caused death there is no doubt whatsoever."

The stories are told over and over in Arizona, New Mexico,

and other states where the lizards roam. Even the Native American tribes in the Southwest were believed by American settlers to have feared the monster's bite; according to one newsman, "The Pima, Apache, Maricopah, and Yuma Indians of the Southwest, who have little fear of the bite of the Mexican centipede or a rattlesnake . . . regard [the Gila monster] as the most to be dreaded of anything that crawls." Legends in the area claim that the lizards are tenacious, refusing to release their jaws once latched on "until the big spirit in the mountains causes a thunder, even if it takes all summer"—a behavior substantiated by several stories told in an article for *The San Francisco Call*: "It is known . . . that it is worse than useless to attempt to force the Gila monster to release its hold." The article also includes the testimony of the U.S. Army captain B. E. Lewis, who claims to have witnessed Gila monsters kill with a hiss. He tells of one of his dogs finding a Gila monster in the yard: "I looked and saw a Gila monster dart forth and hiss in the face of the dog . . . it was dead in two hours. Several persons and I examined the dog's carcass carefully and we could find no evidence of a bite on the beast. I am sure it was

Image accompanying an article from *The San Francisco Call* from 1898, spinning sensationalized stories of the "deadly" Gila monster

something in the hiss or the breath of the Gila monster that killed the dog."

If these stories seem unbelievable to you—that the Gila's jaws are viselike, that they exhale a deadly gas, or that they kill with frightening reliability—you'd be right. No one is quite sure why the lizards developed such a bad rap in the Wild West. The now-threatened Gila monster is a shy and slow creature, in fact, preferring to stay underground in burrows where it's cooler and wetter during the day. And as for its venom, it isn't deadly at all—despite what was written in nineteenth-century newspapers, not a single case of death by Gila envenomation has ever been verified. When you compare it to other venomous species, it's nearly harmless to humans, with venom that is at least one hundred times less potent than known killers. That's toxic enough to hurt like hell, but it's not going to kill anyone easily. And now we know much more than we did back then: the venom also contains a lifesaving compound called exendin, discovered by Eng, that has revolutionized the way doctors treat diabetes.

Eng didn't know much about the lizards or their reputation when he took a chance and ordered the venom, but then again, Eng wasn't a herper. Eng was an endocrinologist at the Veterans Affairs Medical Center in the Bronx, New York. He had just developed new methods to identify unknown hormones with medically useful effects in humans, and was itching to test his protocols. That's when he read about hormones from venomous lizards, discovered by researchers with the National Institutes of Health, that caused enlargement of the pancreas in animal models, a signal that the venom compounds might be overstimulating the organ that produces insulin and other vital hormones. His methods to discover and sequence the venom hormones worked, and he was able to identify a new peptide hormone that no one knew existed. He called his new molecule exendin, to stand for *ex*ocrine, the name for the type of secretion that saliva is, and *en*docrine, a term for hormones.

Eng developed a synthetic version of the Gila monster's venom compound, exendin-4, and sold it to the pharmaceutical company Eli Lilly. The resultant product—Byetta—hit the U.S. pharmaceutical market in 2006. For years it was a billion-dollar drug (until competitors took chunks of the market), and it's not hard to see why: "I went from despair to life—no hope to lots of hope," Rev. John L. Dodson, one of the first wave of patients to be prescribed the new drug, told *The New York Times*. The molecule—dubbed exenatide, to separate the synthetic from the natural molecule taken directly from the lizard's mouth—mimics a hormone, the glucagon-like peptide 1 (or GLP-1, for short), that encourages digestion and the production of insulin. The peptide stimulates insulin release only in the presence of high blood sugar, so unlike with regular insulin injections, there's no accidental hypoglycemia or "insulin coma" from too much insulin. More important, while the human form of GLP-1 is chopped up within minutes in the body (which makes it a terrible pharmaceutical, because it's gone before it can do its job), exenatide lasts for hours.

Byetta was just the beginning. The popularity of the drug ignited a battle between drug companies envious of the peptide's success. A second, similar product, Victoza (liraglutide), was developed by Novo Nordisk and approved by the FDA in 2010. The Byetta alternative was a quick hit; it pulled in more than $1 billion for Novo Nordisk in 2011, its first full year on the market. An extended-release form of exenatide, Bydureon, which is taken weekly instead of daily, was released by Eli Lilly and Amylin Pharmaceuticals in 2012. But a dispute between the companies led to Eli Lilly backing out of the partnership and selling its rights to Amylin, which in turn was purchased by the pharma giants Bristol-Myers Squibb and AstraZeneca, the latter of which now has full rights over the exenatide diabetes duo. Then came Lyxumia (lixisenatide) in 2013, discovered by Zealand Pharma A/S of Denmark and licensed and developed by Sanofi, soon followed to market by GlaxoSmithKline's Tanzeum (albiglutide) and Eli Lilly's Trulicity (dulaglutide) in 2014. All of these very similar drugs continue to be prescribed today in the treatment of diabetes—and none of them would have come to exist if it weren't for the Gila monster.

Further research into exendin-4 suggests that the peptide may be even more of a "miracle drug" than we thought. While scientists with the National Institute on Aging were helping with preclinical testing of exendin-4 in the 1990s, they noticed it didn't just work on the pancreas; it also stimulated neuron growth and kept mature neurons from dying. Based on these initial findings, in 2010 the National Institute on Aging began a human clinical trial to see if the Gila monster peptide could help prevent neurodegeneration when taken daily by people with early-stage Alzheimer's or mild cognitive impairment, a study that is scheduled to finish in 2016. If it goes well, the impact on worldwide health could be enormous. Alzheimer's Disease International estimates that the global cost of dementia (a good portion of which is caused by Alzheimer's) is more than $800 billion—a number that is expected to rise to $2 *trillion* by 2030. Positive results from the trial would pave the way for Byetta to be prescribed for neurodegenerative diseases in addition to diabetes, widening its potential market dramatically.

Byetta actually followed on the heels of one of the most successful pharmaceuticals of all time, captopril. Captopril garnered FDA approval in 1981. Derived from the venom of the jararaca, a Brazilian pit viper, the drug decreases blood pressure by shutting down one of the main pathways for vasoconstriction. Two other snake-derived medicines, Integrilin and Aggrastat, also take advantage of the hemotoxic properties of snakes, acting as anticoagulants. In total, there are six venom-derived drugs on the market today, the result of occasional investigations into the pharmaceutical possibilities hidden in venoms.

"I think the potential is still greater than you would guess from the small number of drugs that have been developed," says Glenn King of the University of Queensland in Brisbane, Australia. Glenn started out as a structural biologist, using complex tools like nuclear magnetic resonance spectrometry, which takes advantage of the magnetic properties of atoms to determine the shape and composition of molecules. Then, one day, a friend called him up and asked for his help in deciphering the structure of a peptide that he'd found in the venom of the Australian funnel web spider. Glenn was so intrigued that he asked for a sample of the venom, and was blown away when he saw how many components there were. "I just thought, 'This is an absolute pharmacological gold mine that nobody's really looked at.'" He's been studying venoms ever since, looking for useful compounds that can serve as everything from insecticides to pharmaceuticals. He's now one of the world's preeminent experts on venoms as therapeutics, having literally edited the book on it.

"In the 1980s and '90s, people weren't saying, 'We should use venoms as a drug source,'" Glenn explained to me. While there were notable discoveries, there was no attempt to systematically screen venoms for potential pharmaceuticals. "Most of it's been serendipitous." But that all changed in the 2000s, when scientists started to look at venoms in a different way. "People started saying, 'Well, actually, these are really complex molecular libraries, and we should start screening them against specific therapeutic targets as a source of drugs.'"

———

We have been toying with the idea for centuries. One of the oldest venom treatments is apitherapy—the medical application of bee venom—which was employed in ancient civilizations in Greece, China, and Egypt. One of the first written accounts of apitherapy can be traced back to Galen, the second-century "father of experimental physiology," who described the use of crushed dead bees mixed with honey as a topical cure for baldness. There are tales that Charlemagne, the eighth-century King of Franks, and Ivan the Terrible (1530–1584), one of Russia's most infamous leaders, used bee stings to treat gout and polyarthritis, respectively. Snakes were also commonly associated with medicine, particularly by the ancient Greeks—so much so that the symbol of two snakes intertwined around a staff with wings, an allusion to a story of the god Hermes, has come to represent the practice of medicine. Traditional Indian medicine, referred to as Ayurveda, frequently employed snake venoms as therapeutics, delivering them on the tip of a needle (a technique called *suchikavoron*) or after a detoxification process (*shodhono*). Stories say that Mithridates VI of Pontus, the so-called poison king (who would go on to become one of the most famous toxinologists of all time) was saved on the battlefield when the coagulative properties of steppe viper venom prevented blood loss from a life-threatening wound—a treatment applied by snake-handling Scythian shamans he'd recruited for his war with Rome.

For the most part, such historical accounts have been considered curious choices made long ago by people who didn't know how bodies worked. At best, these remedies might have been categorized as folk medicines, but largely, they were dismissed alongside other dubious antique medical practices, like bloodletting or trepanation (drilling a hole into one's skull to let out evil spirits).

But then, as the nineteenth century drew to a close, some

doctors and scientists started to realize that the early venom explorers were onto something, and medics again began to dabble in
the use of venoms as therapeutics. A renegade homeopath was
ahead of the curve: John Henry Clarke's *A Dictionary of Practical
Materia Medica* describes the use of several venoms for a diversity
of conditions. Perhaps the most notable examples are his recommendations on the use of cobra venom for pain, a suggestion that
doctors in the first half of the twentieth century followed up with
experiments on people. Low doses of cobra venom worked well for
some subjects with intractable pain and not so well for others.

Other venom tests in modern times have gone well. In addition
to the research above, a recent study found bee stings improved
the symptoms of multiple sclerosis. And even more venoms have
shown promise in animal models, but have yet to be tested in
people, such as the use of snake venom to treat arthritis. Then there
are the multiple assertions, yet to be substantiated, from Bill Haast
and other self-immunizers that low-dose venom injections make
a person healthier overall.

In addition, there are several documented cases where venoms
appear to have treated what doctors failed to. One of the most
incredible stories I have ever heard is that of Ellie Lobel, a woman
who was dying of Lyme disease until she was viciously attacked
by a swarm of Africanized bees.

Lyme disease is caused by spiral-shaped bacteria that can get
injected into our bodies through the venomous bites of deer ticks,
also known as black-legged ticks. Most people who contract the
infection are cured easily with antibiotics if it's discovered early
enough, but in some—for reasons still unclear—the bacteria persist, causing neurodegeneration. Once a smart scientist with a
physics background, Ellie told me that she became incapacitated
by the infection, barely able to stand or form a coherent thought,
let alone live a normal life. She tried everything, but doctor after

doctor and treatment after treatment, she always relapsed. Eventually, she gave up and moved to California to die. She had been there only a few days and was taking a casual stroll when a swarm of bees descended, and Ellie—who'd had a life-threatening allergic reaction to a bee sting as a child—thought that was it. "This is God's way of putting me out of my misery even sooner," she told a friend at the time, refusing treatment for the stings. For days, she was wracked with the worst pain imaginable—but she didn't die. Three years later, she described to me the moment she realized that she might have a future, as the pain finally began to subside. "I thought: I can actually think clearly for the first time in years."

It's impossible to say if the bees truly saved her. But Ellie's case reveals that it's not a completely crazy idea. As she soon discovered, melittin—the most abundant component of bee venoms— is a potent antibiotic. In high doses, it tears holes in bacterial cells, killing them. Though other antibiotics struggle to exterminate the tricky bacteria that cause Lyme, melittin has no trouble with them. If enough melittin made it to the right places, it's *possible* that the compound took out the sinister spirals that were causing Ellie's sickness. There's also evidence that bee and wasp venoms contain components that reverse neurodegeneration and reduce inflammation, both of which are to blame for the worst symptoms in chronic Lyme sufferers. Ellie not only believes that the venom gave her back her life—she's determined to see that bee venoms and melittin are investigated further. She now runs a bee venom cosmetics company, donating some of the venom she obtains for her line of face creams and lotions to be used in cutting-edge research on bee venoms as pharmaceuticals.

Since the early days of venom therapeutics, we have come to know venoms so much more intimately—not just who produces them and what they do, but how they evolve, what makes them tick, and the hundreds to thousands of moving parts that are involved in making them effective toxins. The multitude of different compounds in each potent cocktail, each with a specific molecular target, is what makes venoms such a rich source of potential pharmaceuticals.

At first, scientists studied only the animals that yielded fairly large volumes of venom—snakes, and to a lesser extent, scorpions and spiders. It was a choice born of necessity, as it took a lot of material to run tests on molecules just a few decades ago. "Things have advanced so much in the last ten years," says Glenn King. "You can do screens with tiny micro-amounts of venom which we couldn't do in the past." But even more impressive to him are the advances in the field of genetics that have opened doors for discovery. "We can now genomically look at the toxins in these animals without actually even having to purify the venom. That changes everything."

In particular, venoms have the potential to do what no other pharmaceuticals can. "Whether it's strokes, Alzheimer's, dementias, neurodegenerative conditions, or pain, we really don't have great drugs in these areas," says Ken Winkel, the former head of the Australian Venom Research Unit at the University of Melbourne, "and we've got these little factories of chemical engineers that have a plethora of compounds that may be able to interrupt various pathways or augment certain pathways."

Of course, turning a toxin into a therapeutic isn't a simple process. It can take years, even decades, says Glenn, to go from discovery to market. "There's a lot of things that can go wrong," he notes. "There's isolation of your starting material. There's doing a lot of work understanding structure-activity relationships and making an optimized version. And then doing all the rodent experiments—you have to do rodent experiments to prove that it's efficacious. And it's only at that point you've got any chance really of getting a drug company saying, 'Okay, we're willing to spend the amount of money that's going to be required to put that through a clinical trial.' Then you're looking at probably three to five years to put it through the clinical trials." Only at the end, if a candidate drug makes it through every hoop with flying colors, will it find itself in front of regulatory agencies—the FDA, the European Medicines Agency—to get the final stamp of approval before heading to market. More often than not, the candidate fails—has too many side effects, is too expensive to make or too complex to synthesize, or simply doesn't perform well enough at early stages to be worth the initial investment (which can be hundreds of millions of dollars) to push it through Phase I, II, and III clinical trials.

Even with all those hurdles ahead of them, potential drugs are discovered all the time. A quick scan of venom-related headlines over the past few years shows remarkable strides toward treatments for a diversity of ailments. If you can think of a condition, there's probably a venom-based drug being tested for it. Sea anemone venom tackling autoimmune disorders. Tarantula venom for muscular dystrophy. Centipede venom to cure unrelenting, excruciating pain.

Cancer is a natural target, and venoms have taken aim. There are potential cancer treatments lurking in the venoms of bees, snakes, snails, scorpions, and even mammals. A shrew-derived compound—inelegantly referred to as SOR-C13—concluded a Phase I trial in 2015, making it one step closer to market. (Indeed, some shrews are venomous, with specially grooved teeth to deliver potent toxins to subdue prey.) Meanwhile, the scorpion-sourced "tumor paint" BLZ-100, which helps identify cancers so they can be completely removed, just started a Phase I trial. Doctors hope this "paint" can help guide them during brain tumor surgery in children.

Some of the deadliest infections are also finding their enemies in venom. Scientists recently discovered that one of the major components of bee venom can attack and kill human immunodeficiency virus (HIV), a virus responsible for 1.5 million deaths worldwide every year. Now, they're tweaking the packaging, hoping to create a cure for the currently incurable infection planetwide. Similarly, compounds in snake venoms have shown activity against malaria, a parasite against which modern medical science constantly struggles. If these finds can turn into medicines, they have the potential to save half a billion lives annually and reduce suffering in millions more.

There are less-life-threatening complaints that venoms may alleviate. Got erectile dysfunction? There's a compound in Brazilian wandering spider venom that might be able to straighten that out. Crow's feet? Bee venoms might be better than Botox. There's even a potential spermicide in black widow spider venom—which just seems fitting given how the females of the species are portrayed in popular culture (when you eat your lovers after sex, you get a certain reputation).

Right now, Glenn is working on a painkiller and an epilepsy treatment derived from centipede venoms, among a few other potential drug candidates. "We use arthropod venoms—so, spiders, scorpions, centipedes—because they're all neurotoxic venoms," he explains. "They're a pool of ion-channel-modulating molecules and that's what we're after."

To tackle other diseases—such as heart conditions or blood disorders—animals other than arthropods are a better bet. "If I was after a cardiovascular disease target, then spiders are probably going to be completely useless. Their venoms are not designed to modulate the cardiovascular system of insects, as they don't really have one: it's an open circulation," Glenn notes.

"It's a matter of choosing your venom carefully for the disease target you're after," says Glenn. And with the diversity of venomous species available to choose from, the wealth of potential in venoms is endless. They're so rich in evolved diversity—it's not just the number of venom compounds in a single animal, but all the little tweaks and twists on each theme employed by close relatives, and then all the sundry venomous branches on the tree of life, each with its unique recipe.

Then again, exhausting resources is one of our species' special talents. "We have to look after our venomous biodiversity," Ken Winkel explained to me. "These creatures have developed over millions of years, and it's all too easy to exterminate them without even thinking."

There are species on this planet that we've never seen. They live in lands and seas that no human has ever explored, and they are struggling to survive in a world unknown to us. Though we have not seen or touched them directly, our everyday actions have an impact on their existence. Pollutants that leach from our cities poison their waters. Our trash clutters their landscape with billions of pieces of plastic from which there is no escape. We carelessly alter the planet, without pausing to consider the inevitable impacts on climate. We destroy their homes. And then they are gone, before we even have the chance to meet them.

There are others that we know yet pretend to forget. Rattlesnakes are gleefully rounded up by the tens of thousands every year to supply meat, skins, and sadistic joy to cruel people. The big, watery eyes of the slow loris aren't enough to sway the hearts of poachers who capture the world's only venomous primates to sell as pets, props, or parts. Species that our ancestors revered—snakes, spiders, and scorpions—are treated like inconveniences and intruders, exterminated from their lands with the excuse of safety and security. We are driving the extinction of species at a faster rate than in any other period during life's 3-to-4-billion-year history, and in just a few thousand years, more creatures will die by our hands than by volcanic eruptions, ice ages, or other cataclysmic events.

Every species on this planet tells a story, an evolutionary novel packed with generations upon generations of knowledge. Letting those species disappear is like setting fire to every library on earth. All the information we could ever ask for—the key to understanding life itself—is *right here*. Snakes, spiders, and scorpions, bees and wasps and ants, jellies, fish, urchins, and octopuses, even the bizarre platypus: millions of years of trial and error, data we can never even hope to accrue on our own, will be nothing if we don't preserve the stunning biodiversity of this planet, and by doing so, safeguard biochemical riches.

We can and should conserve venomous species because they are beautiful and wonderful creatures. We can and should protect them because they are integral to the ecosystems they live in, well-oiled parts of an ecological machine that will break down if bits are lost. And the most compelling reason for preserving venomous life is that, through their evolved toxins, venomous animals know more about our bodies than we do. The only way we will ever learn all that these animals have to teach us about ourselves—about *life*—is if we keep them around.

NOTES

1. Masters of Physiology

3 *"Venoms are not accidents"*: Roger A. Caras, "Venomous Animals of the World" (Englewood, NJ: Prentice Hall Trade, 1974), xiii.

3 *"doubt the testimony of my own eyes"*: George Shaw, "*Platypus anatinus*: The Duck-Billed Platypus," *The Naturalist's Miscellany*, vol. 10 (London: F. P. Nodder and Co., 1799), 118.

4 *recognized species of mammals*: Don E. Wilson and DeeAnn M. Reeder, eds., *Mammal Species of the World: A Taxonomic and Geographic Reference*, 3rd ed. (Baltimore: Johns Hopkins University Press, 2005).

5 *a third subcategory of toxic*: David R. Nelsen et al., "Poisons, toxungens, and venoms: Redefining and classifying toxic biological secretions and the organisms that employ them," *Biological Reviews* 89, no. 2 (2014): 450–65.

6–7 *30 milligrams of morphine*: P. J. Fenner et al., "Platypus envenomation—a painful learning experience," *The Medical Journal of Australia* 157 (1992): 829–32.

7 *expressed in the platypus venom gland*: Camilla M. Whittington et al., "Novel venom gene discovery in the platypus," *Genome Biology* 11, no. 9 (2010): R95.

10 *sea snakes switched to eating eggs*: Min Li et al., "Eggs-only diet: Its implications for the toxin profile changes and ecology of the marbled sea snake (*Aipysurus eydouxii*)," *Journal of Molecular Evolution* 60, no. 1 (2005): 81–89.

11 *Mithridates VI of Pontus*: Adrienne Mayor, *The Poison King: The Life and Legend of Mithradates, Rome's Deadliest Enemy* (Princeton, NJ: Princeton University Press, 2011); A. Mayor, "Mithridates of Pontus and His Universal

Antidote," in *History of Toxicology and Environmental Health: Toxicology in Antiquity*, vol. 1, ed. Philip Wexler (Waltham, MA: Academic Press / Elsevier, 2014), 28.

11 *Nicander (roughly 185–135 B.C.)*: *Nicander: The Poems and Poetical Fragments*, ed. and trans. A.S.F. Gow and A. F. Schofield (Cambridge, U.K.: Cambridge University Press, 1953).

11 *Galen (A.D. 131–201)*: Chauncey D. Leake, "Development of knowledge about venoms," in *Venomous Animals and Their Venoms*, vol. I: *Venomous Vertebrates*, ed. Wolfgang Bücherl, Eleanor E. Buckley, and Venancio Deulofeu (New York: Academic Press, 1968), 1.

11 *Francesco Redi (1621?–1697)*: Leake, 8.

11 *the first documented sting was in 1816*: "XXXVI. Extracts from the Minute-Book of the Society. Mar. 18, 1817. Read an Extract of a Letter addressed to the Secretary from Sir John Jamison, F.L.S., dated at Regentville, New South Wales, September 10, 1816," *Transactions of the Linnean Society of London* 12, no. 2 (1818): 584–85.

12 *"lieu dans les serpens venimeux"*: M. H. de Blainville, "Observations sur l'organe appelé Ergot dans l'ornithorinque," *Journal de Physique, de Chimie, d'Histoire Naturelle et des Arts* 84 (1817): 318–20.

12 *"which the spur is conjoined to"*: Anonymous, letter to the editor, *The Sydney Gazette and New South Wales Advertiser*, December 4, 1823, p. 4.

12 *"I should not fear a scratch from one"*: T. Axford, "Notice regarding the Ornithorhynchus," *Edinburgh New Philosophical Journal* 6 (1829): 399–400.

12 *"ignorance of practical natural history"*: Arthur Nicols, *Zoological Notes: On the Structure, Affinities, Habits, and Mental Faculties of Wild and Domestic Animals* (London: L. Upcott Gill, 1883), 122.

13 *Albert Calmette (a protégé of Louis Pasteur)*: Barbara J. Hawgood, "Doctor Albert Calmette 1863–1933: Founder of antivenomous serotherapy and of antituberculous BCG vaccination," *Toxicon* 37, no. 9 (1999): 1241–58.

14 *"Is the platypus venomous?"*: "Poisoned Wounds produced by the Duck-mole (Platypus)," *British Medical Journal* (June 16, 1894): 1332.

14 *first experiment on a live animal*: C. J. Martin and Frank Tidswell, "Observations on the femoral gland of Ornithorhynchus and its secretion; together with an experimental enquiry concerning its supposed toxic action," *Proceedings of the Linnean Society of New South Wales* 9 (1895): 471–500.

14 *"feebly toxic viperine venom"*: C. H. Kellaway and D. H. Le Messurier, "The venom of the platypus (*Ornithorhynchus anatinus*)," *Australian Journal of Experimental Biological and Medical Sciences* 13 (1935): 205–21.

17 *could extract only 100 microliters*: Camilla M. Whittington et al., "Understanding and utilising mammalian venom via a platypus venom transcriptome," *Journal of Proteomics* 72, no. 2 (2009): 155–64.

18 *picked up where Temple-Smith left off*: G. De Plater, R. L. Martin, and P. J. Milburn, "A pharmacological and biochemical investigation of the venom from the platypus (*Ornithorhynchus anatinus*)," *Toxicon* 33, no. 2 (1995): 157–69.

2. Death Becomes Them

21 *Angel Yanagihara stepped into the water*: Angel Yanagihara, e-mail exchanges, December 12, 2012–November 19, 2015.

22 *more than 600 million years ago*: Douglas H. Erwin et al., "The Cambrian conundrum: Early divergence and later ecological success in the early history of animals," *Science* 334 (2011): 1091–97.

22 *deadly venom in less than a second*: T. Holstein and P. Tardent, "An ultrahigh-speed analysis of exocytosis: Nematocyst discharge," *Science* 223 (1984): 830–33.

22 *punches holes in the membranes of cells*: Angel Yanagihara and Ralph V. Shohet, "Cubozoan venom-induced cardiovascular collapse is caused by hyperkalemia and prevented by zinc gluconate in mice," *PLoS ONE* 7, no. 12 (2012): e51368, Figure 4.

22 *to leak potassium and then hemoglobin*: Yanagihara and Shohet.

23 *snake-like proteins and spidery enzymes*: Mahdokht Jouiaei et al., "Firing the sting: Chemically induced discharge of cnidae reveals novel proteins and peptides from box jellyfish (*Chironex fleckeri*) venom," *Toxins* 7, no. 3 (2015): 936–50.

23 *"the most venomous animal"*: "Jellyfish Gone Wild," National Science Foundation, www.nsf.gov/news/special_reports/jellyfish/textonly/locations_australia.jsp.

24 *Water has an LD_{50}*: Val Tech Diagnostics, Inc., "Water: Safety Data Sheet," November 15, 2013, revised September 12, 2014, 4.

24 *1 nanogram per kilogram*: B. Zane Horowitz, "Botulinum toxin," *Critical Care Clinics* 21, no. 4 (2005): 825–39.

25 Chironex fleckeri *0.011 (i.v.)*: Gary J. Calton and Joseph W. Burnett, "Partial purification of *Chironex fleckeri* (sea wasp) venom by immunochromatography with antivenom," *Toxicon* 24, no. 4 (1986): 416–20.

25 Latrodectus mactans *0.90 (s.c.)*: Frank F. S. Daly et al., "Neutralization of *Latrodectus mactans* and *L. hesperus* venom by redback spider (*L. hasseltii*)

antivenom," *Journal of Toxicology: Clinical Toxicology* 39, no. 2 (2001): 119–23.

25 Androctonus crassicauda *0.08 (i.v.)–0.40 (s.c.)*: F. Hassan, "Production of scorpion antivenin," in *Handbook of Natural Toxins*, vol. 2: *Insect Poisons, Allergens, and Other Invertebrate Venoms*, ed. Anthony T. Tu. (New York and Basel: Marcel Dekker, 1984), 577–605.

25 Otostigmus scabricauda *0.6 (i.v.)*: National Institute for Occupational Safety and Health, Registry of Toxic Effects of Chemical Substances (RTECS), December 3, 1998.

25 Lonomia obliqua *9.5 (i.v.)*: 0.19 mg of *Lonomia* bristle extract per 18–20g mouse, A. C. Rocha-Campos et al., "Specific heterologous F(ab')$_2$ antibodies revert blood incoagulability resulting from envenoming by *Lonomia obliqua* caterpillars," *The American Journal of Tropical Medicine and Hygiene* 64 (2001): 283–89.

25 Pogonomyrmex maricopa *0.10 (i.p.)–0.12 (i.v.)*: Patricia J. Schmidt, Wade C. Sherbrooke, and Justin O. Schmidt, "The detoxification of ant (*Pogonomyrmex*) venom by a blood factor in horned lizards (*Phrynosoma*)," *Copeia* no. 3 (1989): 603–607; J. O. Schmidt, "Hymenopteran venoms: Striving towards the ultimate defense against vertebrates," in *Insect Defenses: Adaptive Mechanisms and Strategies of Prey and Predators*, ed. David L. Evans and J. O. Schmidt. (Albany, NY: SUNY Press, 1990), 387–419.

25 Conus geographus *0.001–0.03*: Shigeo Yoshiba, "An estimation of the most dangerous species of cone shell, *Conus (Gastridium) geographus* Linne, 1758, venom's lethal dose in humans," *Japanese Journal of Hygiene* 39 (1984): 565–72; Sébastien Dutertre et al., "Intraspecific variations in *Conus geographus* defence-evoked venom and estimation of the human lethal dose," *Toxicon* 91, no. 1 (2014): 135–44.

25 Tripneustes gratilla *0.05 (i.p.)–0.15 (i.v.)*: Charles Baker Alender, "The venom from the heads of the globiferous pedicellariae of the sea urchin, *Tripneustes gratilla* (Linnaeus)" (Ph.D. dissertation, University of Hawaii, 1964), 87; Bruce W. Halstead, "Current status of marine biotoxicology—An overview," *Clinical Toxicology* 18, no. 1 (1981): 9.

25 Urolophus halleri *28.0 (i.v.)*: Halstead, 12.

25 Synanceia horrida *0.02 (i.p.)–0.3 (i.v.)*: Halstead, 13; H. E. Khoo et al., "Biological activities of *Synanceja horrida* (stonefish) venom," *Natural Toxins* 1, no. 1 (1992): 54–60.

25 Aparasphenodon brunoi *0.16 (i.p.)–>1.6 (s.c.)*: Subcutaneous value calculated to be greater than nonlethal paw injection, documented in C. Wilcox, "Venomous Frogs Are Super-Awesome, but They Are Not Going to Kill

You (I Promise)," *Science Sushi* (Discover Magazine Blogs), August 7, 2015, http://blogs.discovermagazine.com/science-sushi/2015/08/07/venomous -frogs-are-super-awesome-but-they-are-not-going-to-kill-you-i -promise/. Intraperitoneal value given in Carlos Jared et al., "Venomous frogs use heads as weapons," *Current Biology* 25, no. 16 (2015): 2166–70.

25 Oxyuranus microlepidotus *0.025 (s.c.)*: Olga Pudovka Gross and Gus A. Gross, *Management of Snakebites: Study Manual and Guide for Health Care Professionals* (Victoria, BC, Canada: FriesenPress, 2011), 91.

25 Oxyuranus scutellatus *0.013 (i.v.)–0.11 (s.c.)*: Ibid.

25 Crotalus scutulatus *0.03 (i.v.)*: George R. Zug and Carl H. Ernst, *Snakes in Question: The Smithsonian Answer Book* (Washington, D.C.: Smithsonian Institution, 2015).

25 Blarina brevicauda *13.5–21.8 (i.p.)*: Sydney Ellis and Otto Kraver, "Properties of a toxin from the salivary gland of the shrew, *Blarina brevicauda*," *Journal of Pharmacology and Experimental Therapeutics* 114, no. 2 (1955): 127–37.

26 *the coastal taipan is the deadliest*: Gross and Gross, 91.

26 *the same snake falls several slots*: A. J. Broad, S. K. Sutherland, and A. R. Coulter, "The lethality in mice of dangerous Australian and other snake venom," *Toxicon* 17, no. 6 (1979): 661–64.

26 tetrodotoxin, *the main component*: Subcutaneous LD_{50} of 0.0125 mg/kg documented in Qinhui Xu et al., "[Toxicity of tetrodotoxin towards mice and rabbits]" (article in Chinese), *Wei Sheng Yan Jiu* (Journal of Hygiene Research) 32, no. 4 (2003): 371–74.

26 Alatina alata—*have an LD_{50} ranging from 0.005 to 0.025 mg/kg*: Hiroshi Nagai et al., "A novel protein toxin from the deadly box jellyfish (Sea Wasp, Habu-kurage) *Chiropsalmus quadrigatus*," *Bioscience, Biotechnology, and Biochemistry* 66, no. 1 (2002): 97–102.

26 palytoxin, *which, with an LD_{50} of 0.00015 mg/kg*: Halstead, 6.

27 *there are at least sixteen jellies identified*: Lisa-Ann Gershwin, "Two new species of box jellies (Cnidaria: Cubozoa: Carybdeida) from the central coast of Western Australia, both presumed to cause Irukandji syndrome," *Records of the Western Australian Museum* 29 (2014): 10–19.

27 *guinea pigs are ten times more sensitive*: Sergio Bettini, M. Maroli, and Z. Maretić, "Venoms of Theridiidae, genus *Latrodectus*," in *Arthropod Venoms: Handbook of Experimental Pharmacology*, ed. S. Bettini (Berlin and Heidelberg: Springer, 1978), 160.

27 *kills less than 0.5 percent of the people*: Bart J. Currie and Susan P. Jacups, "Prospective study of *Chironex fleckeri* and other box jellyfish stings in the

'Top End' of Australia's Northern Territory," *Medical Journal of Australia* 183, nos. 11/12 (2005): 631.

27 *50 to 60 percent of king cobra bites are fatal*: "*Ophiophagus hannah*," Clinical Toxinology Resources, The University of Adelaide, www.toxinology .com/fusebox.cfm?fuseaction=main.snakes.display&id=SN0048.

27 *about 2 percent of venomous snakebites overall*: Anuradhani Kasturiratne et al., "The global burden of snakebite: A literature analysis and modelling based on regional estimates of envenoming and deaths," *PLoS Medicine* 5, no. 11 (2008): e218.

27–28 *anywhere from 60 to 80 percent*: Kavitha Saravu et al., "Clinical profile, species-specific severity grading, and outcome determinants of snake envenomation: An Indian tertiary care hospital-based prospective study," *Indian Journal of Critical Care Medicine* 16, no. 4 (2012): 187; M. L. Ahuja and G. Singh, "Snake bite in India," in *Venoms*, ed. E. E. Buckley and N. Porges (Washington, D.C.: American Association for the Advancement of Science, 1956), 341–52.

28 Lonomia *moth caterpillars stand out*: Linda Christian Carrijo-Carvalho and Ana Marisa Chudzinski-Tavassi, "The venom of the *Lonomia* caterpillar: An overview," *Toxicon* 49, no. 6 (2007): 741–57.

28 *a case-fatality rate of 20 percent*: Pedro André Kowacs et al., "Fatal intracerebral hemorrhage secondary to *Lonomia obliqua* caterpillar envenoming: Case report," *Arquivos de Neuro-Psiquiatria* 64, no. 4 (2006): 1030–32.

28 *a case-fatality rate of 70 percent*: Centers for Disease Control and Prevention, *Biosafety in Microbiological and Biomedical Laboratories*, 5th ed., Section VIII-G: "Toxin Agents," HHS Publication No. (CDC) 21–1112 (2009): 276.

29 *kill tens of thousands of people every year*: Ian D. Simpson and Robert L. Norris, "Snakes of medical importance in India: Is the concept of the 'Big 4' still relevant and useful?" *Wilderness and Environmental Medicine* 18, no. 1 (2007): 2–9.

29 *Hindu code of laws known as the Gentoo Code*: Nathaniel Brassey Halhead, *A Code of Gentoo Laws, or, Ordinations of the Pundits, from a Persian Translation, Made from the Original, Written in the Shanscrit Language* (London, 1776).

29 *the Vish Kanya—legendary young women assassins*: Vipul Namdeorao Ambade, Jaydeo Laxman Borkar, and Satin Kalidas Meshram, "Homicide by direct snake bite: A case of contract killing," *Medicine, Science and the Law* 52, no. 1 (2012): 40–43.

30 *Glenn Summerford, a snake-handling preacher*: Thomas G. Burton, *The Serpent and the Spirit: Glenn Summerford's Story* (Knoxville: The University of Tennessee Press, 2004).

30 *According to Darlene, in October 1991*: Burton, 5–15.

31 *Glenn, of course, told a very different story*: Burton, 125–40.

31 *Robert "Rattlesnake" James made history*: Gerald F. Uelmen, "Memorable Murder Trials of Los Angeles," *Los Angeles Lawyer* 4 (March 1981): 21.

31 *a man paid a kidnapper*: Ambade et al., 40–41.

32 *paraded an effigy of her body*: W. Ralph Johnson, "A quean, a great queen? Cleopatra and the politics of misrepresentation," *Arion* 6, no. 3 (1967): 387–402.

32 *to give a person spiritual immortality*: Michael Grant, *Cleopatra* (Edison, NJ: Castle Books, 1972; repr. 2004), 216–28.

32 *a humane method of execution*: François P. Retief and Louise Cilliers, "The death of Cleopatra," *Acta Theologica* 26, no. 2, Supplementum 7 (2005): 79–88.

32 *infamous Viking conqueror Ragnar Lothbrok*: Robert W. Rix, "The afterlife of a death song: Reception of Ragnar Lodbrog's poem in Britain until the end of the eighteenth century," *Studia Neophilologica* 81, no. 1 (2009): 53–68.

32 *killed worldwide by snakebites every year*: Kasturiratne et al., e218.

33 *even the serpent in Eden*: Henry Ansgar Kelly, "The metamorphoses of the Eden serpent during the Middle Ages and Renaissance," *Viator* 2 (1972): 301–28.

33 *associated with intelligence and elegance*: Balaji Mundkur et al., "The cult of the serpent in the Americas: Its Asian background [and comments and reply]," *Current Anthropology* 17, no. 3 (1976): 429–55.

33 *the real driver of acute vision*: Lynne A. Isbell, *The Fruit, the Tree, and the Serpent: Why We See So Well* (Cambridge, MA: Harvard University Press, 2009).

34 *in these New World monkeys*: Isbell, 104–106.

35 *are innately afraid of snakes*: Quan Van Le et al., "Pulvinar neurons reveal neurobiological evidence of past selection for rapid detection of snakes," *PNAS* 110, no. 47 (2013): 19000–19005; Judy S. DeLoache and Vanessa LoBue, "The narrow fellow in the grass: Human infants associate snakes and fear," *Developmental Science* 12, no. 1 (2009): 201–207.

35 *where we can't spot spiders*: Sandra C. Soares et al., "The hidden snake in the grass: Superior detection of snakes in challenging attentional conditions," *PLoS ONE* 9, no. 12 (2014): e114724.

36 *nonthreatening shapes like mushrooms or flowers*: Arne Öhman and Joaquim J. F. Soares, "'Unconscious anxiety': Phobic responses to masked stimuli," *Journal of Abnormal Psychology* 103, no. 2 (1994): 231–40.

36 *Isbell hypothesizes that the switch to bipedalism*: Isbell, 145–53.

36 *kill more people every year in the United States*: Ricky L. Langley, "Animal-related fatalities in the United States—an update," *Wilderness and Environmental Medicine* 16, no. 2 (2005): 67–74.

37 *their hematophagous (or blood-feeding) lifestyle*: J.M.C. Ribeiro, "Role of saliva in blood-feeding by arthropods," *Annual Review of Entomology* 32 (1987): 463–78.

38 *Malaria claims more than 600,000*: World Health Organization, Global Health Observatory (GHO) data: "Number of malaria deaths: Estimated deaths, 2012," www.who.int/gho/malaria/epidemic/deaths/en/.

38 *30,000 from yellow fever*: World Health Organization, fact sheet no. 100, "Yellow Fever," updated March 2014, www.who.int/mediacentre/factsheets/fs100/en/.

38 *12,000 from dengue*: World Health Organization, fact sheet no. 117, "Dengue and Severe Dengue," updated May 2015, www.who.int/mediacentre/factsheets/fs117/en/.

38 *20,000 from Japanese encephalitis*: World Health Organization, fact sheet no. 386, "Japanese Encephalitis," December 2015, www.who.int/mediacentre/factsheets/fs386/en/.

38 *40 million people disfigured*: World Health Organization, fact sheet no. 102, "Lymphatic Filariasis," updated May 2015, www.who.int/mediacentre/factsheets/fs102/en/.

38 *the esteemed journal* Nature *asked*: Janet Fang, "Ecology: A world without mosquitoes," *Nature* 466 (2010): 432–34.

38 *from every single caribou in a herd*: Ibid., 433.

39 *wiped out at elevations where mosquitoes flourish*: Richard E. Warner, "The role of introduced diseases in the extinction of the endemic Hawaiian avifauna," *The Condor* 70 (1968): 101–20.

3. Of Mongeese and Men

41 *"horror filtered through my mind"*: Joel La Rocque, "Self Immunization—A Dangerous Road to Travel," *ezine articles*, September 18, 2009, http://ezinearticles.com/?Self-Immunization—A-Dangerous-Road-To-Travel&id=2947421.

48 *from three to six liters of blood*: World Health Organization, "WHO Guidelines for the Production Control and Regulation of Snake Antivenom Immunoglobulins" (Geneva, Switzerland: WHO Press, 2010), www.who.int/bloodproducts/snake_antivenoms/SnakeAntivenomGuideline.pdf.

49 *43 to 81 percent of snakebite victims*: I. B. Gawarammana et al., "Parallel infusion of hydrocortisone±chlorpheniramine bolus injection to prevent

acute adverse reactions to antivenom for snakebites," *Medical Journal of Australia* 180 (2004): 20–23; C. A. Ariaratnam et al., "An open, randomized comparative trial of two antivenoms for the treatment of envenoming by Sri Lankan Russell's viper (*Daboia russelii russelii*)," *Transactions of the Royal Society of Tropical Medicine and Hygiene* 95, no. 1 (2001): 74–80; A. P. Premawardhena et al., "Low dose subcutaneous adrenaline to prevent acute adverse reactions to antivenom serum in people bitten by snakes: Randomised, placebo controlled trial," *British Medical Journal* 318 (1999): 1041–43.

51 *forty to eighty times the dose of viper venom*: H. Moussatché and J. Perales, "Factors underlying the natural resistance of animals against snake venoms," *Memórias do Instituto Oswaldo Cruz* 84, Suppl. IV (1989): 391–94.

51 *able to withstand three to twenty times the amount*: Ashlee H. Rowe and Matthew P. Rowe, "Physiological resistance of grasshopper mice (*Onychomys* spp.) to Arizona bark scorpion (*Centruroides exilicauda*) venom," *Toxicon* 52 (2008): 597–605.

51 *from the vipers it feeds upon*: Sameh Darawshi, "The ecology of the Short-toed Eagle (*Circaetus gallicus*) in the Judean Slopes, Israel" (graduate thesis, The Hebrew University of Jerusalem, 2013), www.rufford.org/files/sameh_darawshi_RSG_Final.pdf.

52 *The little lizard can survive*: Eliahu Zlotkin et al., "Predatory behaviour of gekkonid lizards, *Ptyodactylus* spp., towards the scorpion *Leiurus quinquestriatus hebraeus*, and their tolerance of its venom," *Journal of Natural History* 37, no. 5 (2003): 641–46.

52 *the most potent venom in the Hymenoptera*: W. L. Meyer, "Most Toxic Insect Venom," Book of Insect Records, University of Florida, May 1, 1996.

52 *fifteen hundred times that of mice*: Schmidt, Sherbrooke, and Schmidt, 606.

52 *by Texas scientists in the 1970s*: John C. Perez, Willis C. Haws, and Curtis H. Hatch, "Resistance of woodrats (*Neotoma micropus*) to *Crotalus atrox* venom," *Toxicon* 16, no. 2 (1978): 198–200.

52 *to purify the serum compound*: Vivian E. Garcia and John C. Perez, "The purification and characterization of an antihemorrhagic factor in woodrat (*Neotoma micropus*) serum," *Toxicon* 22, no. 1 (1984): 129–38.

52 *few to no signs of trouble*: Harold Heatwole and Judy Powell, "Resistance of eels (*Gymnothorax*) to the venom of sea kraits (*Laticauda colubrina*): A test of coevolution," *Toxicon* 36, no. 4 (1998): 619–25.

52 *other venoms within the same family*: Michael Ovadia and E. Kochva, "Neutralization of Viperidae and Elapidae snake venoms by sera of different animals," *Toxicon* 15, no. 6 (1977): 541–47.

53 *the killer activity of cottonmouth venom*: Dorothy E. Bonnett and Sheldon I. Guttman, "Inhibition of moccasin (*Agkistrodon piscivoris*) venom proteolytic activity by the serum of the Florida king snake (*Lampropeltis getulus floridana*)," *Toxicon* 9, no. 4 (1971): 417–25.

53 *against cobras and other elapid snakes*: Robert S. Voss and Sharon A. Jansa, "Snake-venom resistance as a mammalian trophic adaptation: Lessons from didelphid marsupials," *Biological Reviews* 87, no. 4 (2012): 822–37; Robert M. Werner and James A. Vick, "Resistance of the opossum (*Didelphis virginiana*) to envenomation by snakes of the family Crotalidae," *Toxicon* 15, no. 1 (1977): 29–32.

53 *given* thirteen times that amount: Avner Bdolah et al., "Resistance of the Egyptian mongoose to sarafotoxins," *Toxicon* 35, no. 8 (1997): 1251–61.

53 *is innate and cannot be shared*: Ovadia and Kochva, "Neutralization of Viperidae and Elapidae snake venoms."

54 *the snakes' potent receptor-targeting toxins*: Dora Barchan et al., "How the mongoose can fight the snake: The binding site of the mongoose acetylcholine receptor," *PNAS* 89 (1992): 7717–21.

54 *independently at least four times*: Danielle H. Drabeck, Antony M. Dean, and Sharon A. Jansa, "Why the honey badger don't care: Convergent evolution of venom-targeted nicotinic acetylcholine receptors in mammals that survive venomous snake bites," *Toxicon* 99 (2015): 68–72.

54 *circulating in their blood*: Alexis Rodriguez-Acosta, Irma Aguilar, and Maria E. Giron, "Antivenom activity of opossum (*Didelphis marsupialis*) serum fraction," *Toxicon* 33, no. 1 (1995): 95–98; Jonas Perales et al., "Neutralization of the oedematogenic activity of *Bothrops jararaca* venom on the mouse paw by an antibothropic fraction isolated from opossum (*Didelphis marsupialis*) serum," *Inflammation and Immunomodulation: Agents and Actions* 37 (1992): 250–59; Ana G. C. Neves-Ferreira et al., "Isolation and characterization of DM40 and DM43, two snake venom metalloproteinase inhibitors from *Didelphis marsupialis* serum," *Biochimica et Biophysica Acta—General Subjects* 1474, no. 3 (2000): 309–20.

55 *to their young through their milk*: P. B. Jurgilas et al., "Detection of an antibothropic fraction in opossum (*Didelphis marsupialis*) milk that neutralizes *Bothrops jararaca* venom," *Toxicon* 37, no. 1 (1999): 167–72.

55 *special components in their blood*: Cynthia A. de Wit and Björn R. Weström, "Venom resistance in the hedgehog, *Erinaceus europaeus*: Purification and identification of macroglobulin inhibitors as plasma antihemorrhagic factors," *Toxicon* 25, no. 3 (1987): 315–23; Tamotsu Omori-Satoh, Yoshio

Yamakawa, and Dietrich Mebs, "The antihemorrhagic factor, erinacin, from the European hedgehog (*Erinaceus europaeus*), a metalloprotease inhibitor of large molecular size possessing ficolin/opsonin P35 lectin domains," *Toxicon* 38, no. 11 (2000): 1561–80.

55 *share a lot of similarities*: G. B. Domont, J. Perales, and H. Moussatché, "Natural anti-snake venom proteins," *Toxicon* 29, no. 10 (1991): 1183–94.

55 *serum proteins from the predators*: Sharon A. Jansa and Robert S. Voss, "Adaptive evolution of the venom-targeted vWF protein in opossums that eat pitvipers," *PLoS ONE* 6 (2011): e20997.

56 *written songs with Slash*: Nuxx, "Steve Ludwin," www.nuxx.com/section .php?id=sl.

56 *a date with Courtney Love*: Steve Ludwin, "The Day Kurt Cobain Threatened to Kill My Girlfriend," *Noisey: Music by Vice*, July 14, 2014, http:// noisey.vice.com/en_uk/blog/the-day-kurt-cobain-threatened-to-kill -my-girlfriend-steve-ludwin.

56 *Steve explained to me*: Steve Ludwin, phone interview, January 21, 2015.

57 *Greek* herpetó, *or "creeping thing"*: Gordon Gordh and David Headrick, *A Dictionary of Entomology*, 2nd ed. (Wallingford, U.K., and Cambridge, MA: CABI International, 2011), 625.

57 *Bill Haast, the director of the Miami Serpentarium*: Nancy Haast, The Official Website of W. E. "Bill" Haast, www.billhaast.com/.

57 *began with cobra venom in 1948*: N. Haast, "Snakebites and Immunity," www.billhaast.com/serpentarium/immunization_snakebites.html.

57 *bitten more than* 170 *times*: N. Haast, "Snakebites and Immunity."

58 *When he was eighty-eight years old*: VisualSOLUTIONSMedia, "Bill Haast, Snake Man: An American Original," *YouTube*, June 23, 2011, www.youtube .com/watch?v=hDAaXQJ9BtU.

59–60 *he conducted an online discussion forum*: Steve Ludwin, "IAmA guy who's been injecting deadly snake venom into myself for 20 years. AMA [Ask Me Anything]," *Reddit*, January 29, 2013, www.reddit.com/r/IAmA /comments/17hzhk/iama_guy_whos_been_injecting_deadly_snake _venom/.

62 *isn't being paid a penny for his contributions*: Andreas H. Laustsen et al., "Snake venomics of monocled cobra (*Naja kaouthia*) and investigation of human IgG response against venom toxins." *Toxicon* 99 (2015): 23–35.

63 *may play a role in the fight against parasites*: R. G. Bell, "IgE, allergies and helminth parasites: A new perspective on an old conundrum," *Immunology and Cell Biology* 74 (1996): 337–45.

63 *the elusive scientist Margie Profet*: Margie Profet, "The function of allergy:

Immunological defense against toxins," *Quarterly Review of Biology* 66, no. 1 (1991): 23–62.

66 *small doses of bee venom*: Thomas Marichal et al., "A beneficial role for immunoglobulin E in host defense against honeybee venom," *Immunity* 39, no. 5 (2013): 963–75; Dario A. Gutierrez and Hans-Reimer Rodewald, "A sting in the tale of Th2 immunity," *Immunity* 39, no. 5 (2013): 803–805.

4. To the Pain

68 *a three-inch nail in your heel*: Justin O. Schmidt, *The Sting of the Wild* (Baltimore, MD: Johns Hopkins University Press, 2016), 221–30.

69 *gain respect and leadership*: Vidal Haddad Junior, João Luiz Costa Cardoso, and Roberto Henrique Pinto Moraes, "Description of an injury in a human caused by a false tocandira (*Dinoponera gigantea*, Perty, 1833) with a revision on folkloric, pharmacological and clinical aspects of the giant ants of the genera *Paraponera* and *Dinoponera* (sub-family Ponerinae)," *Revista do Instituto de Medicina Tropical de São Paulo* 47, no. 4 (2005): 235–38.

70 *after collapsing from the unrelenting agony*: Hamish and Andy, "The worst pain known to man," *YouTube*, August 5, 2014, www.youtube.com/watch?v=it0V7xv9qu0.

70 *for hours after the stings*: National Geographic, "Wearing a Glove of Venomous Ants," *YouTube*, March 3, 2011, www.youtube.com/watch?v=XEWmynRcEEQ.

70 *to withstand the sting of the bullet ant*: Steve Backshall, "Bitten by the Amazon," *The Sunday Times*, January 6, 2008, www.thesundaytimes.co.uk/sto/travel/Holidays/Wildlife/article77936.ece.

77 *an unfortunate encounter with a lionfish*: Heinz Steinitz, "Observations on *Pterois miles* (L.) and its venom," *Copeia* no. 2 (1959): 159–61.

78 *not just cryptic—they're "repulsively ugly"*: Albert Calmette, *Venoms: Venomous Animals and Antivenomous Serum-Therapeutics*, trans. Ernest E. Austen (New York: William Wood and Company, 1908), 290.

78 *"get him to shore without drowning"*: J.L.B. Smith, "A case of poisoning by the stonefish, *Synanceja verrucosa*," *Copeia* no. 3 (1951): 207–10.

78 *"may become almost demented, and . . . may die"*: N. K. Cooper, "Stone fish and stingrays—some notes on the injuries that they cause to man," *Journal of the Royal Army Medical Corps* 137, no. 3 (1991): 136–40.

79 *his death would come from the sea*: Edmund D. Cressman, "Beyond the Sunset," *Classical Journal* 27, no. 9 (1932): 669–74.

79 *died at the age of forty-four*: Rene Lynch, "'Crocodile Hunter' cameraman: Footage of Steve Irwin death is private," *Los Angeles Times*, March 10, 2014, www.latimes.com/nation/la-sh-crocodile-hunter-steve-irwins-last -words-im-dying-20140310-story.html.

85 *the added metabolic cost of a baby*: K. Melzer et al., "Pregnancy-related changes in activity energy expenditure and resting metabolic rate in Switzerland," *European Journal of Clinical Nutrition* 63, no. 10 (2009): 1185–91.

85 *by 11 percent for three days*: Marshall D. McCue, "Cost of producing venom in three North American pitviper species," *Copeia* no. 4 (2006): 818–25.

85 *the first three days of venom production*: A.F.V. Pintor, A. K. Krockenberger, and J. E. Seymour, "Costs of venom production in the common death adder (*Acanthophis antarcticus*)," *Toxicon* 56, no. 6 (2010): 1035–42.

85 *by less than 10 percent on average*: Heidi K. Byrne and Jack H. Wilmore, "The effects of a 20-week exercise training program on resting metabolic rate in previously sedentary, moderately obese women," *The International Journal of Sport Nutrition and Exercise Metabolism* 11 (2001): 15–31; Jeffrey T. Lemmer et al., "Effect of strength training on resting metabolic rate and physical activity: age and gender comparisons," *Medicine and Science in Sports and Exercise* 33 (2001): 532–41; and J. C. Aristizabal et al., "Effect of resistance training on resting metabolic rate and its estimation by a dual-energy X-ray absorptiometry metabolic map," *European Journal of Clinical Nutrition* 69 (2014): 831–36.

86 *eight days when replenishing venom*: Zia Nisani, Stephen G. Dunbar, and William K. Hayes, "Cost of venom regeneration in *Parabuthus transvaalicus* (Arachnida: Buthidae)," *Comparative Biochemistry and Physiology Part A: Molecular and Integrative Physiology* 147, no. 2 (2007): 509–13; Nisani et al., "Investigating the chemical profile of regenerated scorpion (*Parabuthus transvaalicus*) venom in relation to metabolic cost and toxicity," *Toxicon* 60, no. 3 (2012): 315–23.

86 *isn't required or won't be effective*: David Morgenstern and Glenn F. King, "The venom optimization hypothesis revisited," *Toxicon* 63 (2013): 120–28.

86 *20 to 50 percent of the bites are dry*: "Fortunately, 50% of bites by venomous snakes are 'dry bites' that result in negligible envenomation," Syed Moied Ahmed et al., "Emergency treatment of a snake bite: Pearls from literature," *Journal of Emergencies, Trauma and Shock* 1, no. 2 (2008): 97–105; "one in every four," Kasturiratne et al., e218.

92 *"and respiratory distress may occur"*: Ming-Ling Wu et al., "Sea-urchin enven-omation," *Veterinary and Human Toxicology* 45, no. 6 (2003): 307–309.

93 *tend to be simpler in composition*: Nicholas R. Casewell et al., "Complex cocktails: The evolutionary novelty of venoms," *Trends in Ecology and Evolution* 28, no. 4 (2013): 219–29.

5. Bleed It Out

97 *150,000 to 350,000 per microliter of blood*: J. N. George, "Platelets," Platelets on the Web, April 6, 2005, www.ouhsc.edu/platelets/platelets /platelets%20intro.html.

100 *Lopap, a 185-amino-acid prothrombin activator*: Cleyson V. Reis et al., "Lopap, a prothrombin activator from Lonomia obliqua belonging to the lipocalin family: Recombinant production, biochemical character-ization and structure-function insights," *Biochemistry Journal* 398 (2006): 295–302.

100 *Losac (*Lonomia obliqua *Stuart factor activator)*: Miryam Paola Alvarez-Flores et al., "Losac, the first hemolin that exhibits procoagulant activity through selective factor X proteolytic activation," *Journal of Biological Chemistry* 286 (2011): 6918–28.

105 *isolated by venom scientists from leeches*: Michel Salzet, "Anticoagulants and inhibitors of platelet aggregation derived from leeches," *FEBS Letters* 492, no. 3 (2001): 187–92.

109 *"on their teeth, cultivating bacteria"*: Tracey Franchi, "Fear of Komodo dragon bacteria wrapped in myth," UQ News, University of Queensland, June 25, 2013, www.uq.edu.au/news/article/2013/06/fear-of-komodo -dragon-bacteria-wrapped-myth.

109 *share the same venom genes*: Bryan G. Fry et al., "Early evolution of the venom system in lizards and snakes," *Nature* 439 (2006): 584–88.

109 *do indeed have venom glands*: Bryan G. Fry et al., "A central role for venom in predation by *Varanus komodoensis* (Komodo Dragon) and the extinct giant *Varanus (Megalania) priscus*," *PNAS* 106, no. 22 (2009): 8969–74.

109 *"incorrect or falsely misleading"*: Kurt Schwenk, quoted in Carl Zimmer, "Chemicals in Dragon's Glands Stir Venom Debate," *The New York Times*, May 19, 2009, www.nytimes.com/2009/05/19/science/19komo.html.

109–110 *more samples and better techniques*: Ellie J. C. Goldstein et al., "Anaerobic and aerobic bacteriology of the saliva and gingiva from 16 captive Komodo dragons (*Varanus komodoensis*): New implications for the 'bacteria as venom' model," *Journal of Zoo and Wildlife Medicine* 44, no. 2 (2013): 262–72.

110 *didn't find the pathogenic species*: The authors went on to explain where the earlier research had gone wrong: of the fifty-four species that previous research claimed to be "potentially pathogenic," thirty-three are actually common microbes and "unlikely to be the cause of rapid fatal infection when present in a wound." None of the species found were virulent enough to cause such rapid death. Bryan and his team didn't find the species the previous team had pointed to in the original paper as the probable cause of sepsis (a species that, the authors noted, was found only in 5 percent of the dragons studied in the first place). The authors also pointed out that the earlier researchers were at a disadvantage, as they had to identify bacteria "without the advantage of molecular methods."

110 *"Having gotten septicemia in Flores"*: Bryan G. Fry, Facebook comment, June 26, 2013.

111 *extremely complex reptile venom gland*: Fry et al., "A central role for venom in predation."

113 *a group of lost divers barely survived*: Richard Edwards, "Stranded divers had to fight off Komodo dragons to survive," *The Telegraph*, June 8, 2008, www.telegraph.co.uk/news/worldnews/asia/indonesia/2095835/Stranded -divers-had-to-fight-off-Komodo-dragons-to-survive.html.

117 *the majority of venomous bites*: Rafael Otero-Patiño, "Epidemiological, clinical and therapeutic aspects of *Bothrops asper* bites," *Toxicon* 54, no. 7 (2009): 998–1011.

117 *a platelet-aggregating compound called aspercetin*: Alexandra Rucavado et al., "Characterization of aspercetin, a platelet aggregating component from the venom of the snake *Bothrops asper* which induces thrombocytopenia and potentiates metalloproteinase-induced hemorrhage," *Thrombosis and Haemostasis* 85 (2001): 710–15.

118 *half of what it was when they were together*: Gadi Borkow, José María Gutiér-rez, and Michael Ovadia, "Isolation and characterization of synergistic hemorrhagins from the venom of the snake *Bothrops asper*," *Toxicon* 31, no. 9 (1993): 1137–50.

118 *in other species of snakes*: P. E. Bougis, P. Marchot, and H. Rochat, "*In vivo* synergy of cardiotoxin and phospholipase A_2 from the elapid snake *Naja mossambica mossambica*," *Toxicon* 25, no. 4 (1987): 427–31.

118 *as well as in bees and hornets*: Miriam Kolko et al., "Synergy by secretory phospholipase A_2 and glutamate on inducing cell death and sustained

arachidonic acid metabolic changes in primary cortical neuronal cultures," *Journal of Biological Chemistry* 271 (1996): 32722–28; C.-L. Ho and L.-L. Hwang, "Structure and biological activities of a new mastoparan isolated from the venom of the hornet *Vespa basalis*," *Biochemical Journal* 274, part 2 (1991): 453–56.

6. All the Better to Eat You With

120 *"certain destruction to her enemies"*: Benjamin Franklin, quoted in *America's Founding Fathers: Their Uncommon Wisdom and Wit*, ed. Bill Adler (Lanham, MD: Taylor Trade Publishing, 2003), 4–8.

124 *no antibodies to inhibit them*: José María Gutiérrez et al., "Experimental pathology of local tissue damage induced by *Bothrops asper* snake venom," *Toxicon* 54, no. 7 (2009): 958–75.

124 *by mechanisms yet unknown*: José María Gutiérrez and Alexandra Rucavado, "Snake venom metalloproteinases: Their role in the pathogenesis of local tissue damage," *Biochimie* 82, no. 9 (2000): 841–50.

125 *that rush to the wound*: Catarina Teixeira et al., "Inflammation induced by *Bothrops asper* venom," *Toxicon* 54, no. 1 (2009): 988–97.

125 *necrosis from snake venoms is greatly reduced*: Gavin David Laing et al., "Inflammatory pathogenesis of snake venom metalloproteinase-induced skin necrosis," *European Journal of Immunology* 33, no. 12 (2003): 3458–63.

125 *mast cells, to release histamine*: Hui-Fen Chiu, Ing-Jun Chen, and Che-Ming Teng, "Edema formation and degranulation of mast cells by a basic phospholipase A_2 purified from *Trimeresurus mucrosquamatus* snake venom," *Toxicon* 27, no. 1 (1989): 115–25.

126 *wasps and their kin*: Uğur Koçer et al., "Skin and soft tissue necrosis following hymenoptera sting," *Journal of Cutaneous Medicine and Surgery: Incorporating Medical and Surgical Dermatology* 7, no. 2 (2003): 133–35.

126 *sometimes cause large lesions*: Peter Barss, "Wound necrosis caused by the venom of stingrays. Pathological findings and surgical management," *Medical Journal of Australia* 141, nos. 12–13 (1984): 854–55.

128 *liquefication—"liquefactive necrosis"*: M. H. Appel et al., "Insights into brown spider and loxoscelism," *Invertebrate Survival Journal* 2 (2005): 152–58.

128 *in the rarest of cases, death*: David L. Swanson and Richard S. Vetter, "Loxoscelism," *Clinics in Dermatology* 24, no. 3 (2006): 213–21.

129 *reduces the dermonecrotic activity*: Patrícia Guilherme, Irene Fernandes, and Katia Cristina Barbaro, "Neutralization of dermonecrotic and lethal

activities and differences among 32–35 kDa toxins of medically important *Loxosceles* spider venoms in Brazil revealed by monoclonal antibodies," *Toxicon* 39, no. 9 (2001): 1333–42.

129 *Sicariids . . . that's it. No other spiders*: Greta J. Binford and Michael A. Wells, "The phylogenetic distribution of sphingomyelinase D activity in venoms of Haplogyne spiders," *Comparative Biochemistry and Physiology Part B: Biochemistry and Molecular Biology* 135, no. 1 (2003): 25–33.

129 *developed their potent necrotic enzyme*: G. J. Binford, Matthew H. J. Cordes, and M. A. Wells, "Sphingomyelinase D from venoms of *Loxosceles* spiders: Evolutionary insights from cDNA sequences and gene structure," *Toxicon* 45, no. 5 (2005): 547–60.

130 *rarely are our open sores the work of tenacious arachnids*: Richard S. Vetter, "Spiders of the genus *Loxosceles* (Araneae, Sicariidae: A review of biological, medical and psychological aspects regarding envenomations," *The Journal of Arachnology* 36 (2008): 150–63.

130 *inflict a necrotic bite with any regularity*: Swanson and Vetter, 215.

130 *venom is almost twice as potent*: Kátia C. de Oliveira et al., "Variations in *Loxosceles* spider venom composition and toxicity contribute to the severity of envenomation," *Toxicon* 45, no. 4 (2005): 421–29.

130 *rid the house of the potential threat*: Richard S. Vetter and Diane K. Barger, "An infestation of 2,055 brown recluse spiders (Araneae: Sicariidae) and no envenomations in a Kansas home: Implications for bite diagnoses in nonendemic areas," *Journal of Medical Entomology* 39, no. 6 (2002): 948–51.

131 *more than 85 percent were bacterial infections*: Jeffrey Ross Suchard, "'Spider bite' lesions are usually diagnosed as skin and soft-tissue infections," *Journal of Emergency Medicine* 41, no. 5 (2011): 473–81.

131 *people who thought they had spider bites*: Tamara J. Dominguez, "It's not a spider bite, it's community-acquired methicillin-resistant *Staphylococcus aureus*," *Journal of the American Board of Family Medicine* 17, no. 3 (2004): 220–26.

132 *scholars believe the author was Ben Franklin*: Walter Isaacson, *Benjamin Franklin: An American Life* (New York: Simon & Schuster, 2003), 305.

133 *Approximately sixty thousand protein families*: Victor Kunin et al., "Myriads of protein families, and still counting," *Genome Biology* 4, no. 2 (2003): 401.

134 *factors that set the venomous proteins apart*: Bryan G. Fry et al., "The toxicogenomic multiverse: Convergent recruitment of proteins into animal venoms," *Annual Review of Genomics and Human Genetics* 10 (2009): 483–511.

7. Don't Move

137 *"a slumberous lethargy brings life's end"*: Quoted in Peter K. Knoefel and Madeline C. Covi, *Hellenistic Treatise on Poisonous Animals (The Theriaca of Nicander of Colophon: A Contribution to the History of Toxicology)* (Lewiston, NY: Edwin Mellen Press, 1991), 99.

138 *at Suttons Beach in Queensland*: Elena Cavazzoni et al., "Blue-ringed octopus (*Hapalochlaena* sp.) envenomation of a 4-year-old boy: A case report," *Clinical Toxicology* 46, no. 8 (2008): 760–61.

138 *his legs were all floppy*: "Boy bitten by octopus," *Gold Coast Bulletin*, October 9, 2006.

138 *quickly killed a full-grown man*: H. Mabbet, "Death of a Skin Diver," *Skin Diving and Spearfishing Digest*, December 1954, 13, 17.

139 *deaths were considered aberrations*: Bruce W. Halstead, *Poisonous and Venomous Marine Animals of the World*, vol. I: *Invertebrates* (Washington, D.C.: U.S. Government Printing Office, 1965), 742–43.

139 *absence of knowledge of its chemical composition*: Shirley E. Freeman and R. J. Turner, "Maculotoxin, a potent toxin secreted by *Octopus maculosus* Hoyle," *Toxicology and Applied Pharmacology* 16, no. 3 (1970): 681–90.

139 *pufferfishes: the infamous tetrodotoxin*: D. D. Sheumack et al., "Maculotoxin: A neurotoxin from the venom glands of the octopus *Hapalochlaena maculosa* identified as tetrodotoxin," *Science* 199 (1978): 188–89.

140 *among the deadliest compounds known to man*: Toshio Narahashi, "Tetrodotoxin: A brief history," *Proceedings of the Japan Academy, Series B, Physical and Biological Sciences* 84, no. 5 (2008): 147–54.

141 *mechanoreceptors just beneath my skin's outermost layer*: Dale Purves et al., "Mechanoreceptors Specialized to Receive Tactile Information," in *Neuroscience*, 2nd ed., D. Purves et al., eds. (Sunderland, MA: Sinauer Associates, 2001).

141 *opening force-sensitive ion channels*: Ellen A. Lumpkin, Kara L. Marshall, and Aislyn M. Nelson, "The cell biology of touch," *The Journal of Cell Biology* 191, no. 2 (2010): 237–48.

143 *sodium channels that it isn't effective against*: Chong Hyun Lee and Peter C. Ruben, "Interaction between voltage-gated sodium channels and the neurotoxin, tetrodotoxin," *Channels* 2, no. 6 (2008): 407–12; Narahashi, 152–53.

143 *resistant or immune to its effects*: Toshio Saito et al., "Tetrodotoxin as a biological defense agent for puffers," *Nippon Suisan Gakkaishi* 51, no. 7 (1985): 1175–80; Tamao Noguchi and Osamu Arakawa, "Tetrodotoxin—distribution and

accumulation in aquatic organisms, and cases of human intoxication," *Marine Drugs* 6, no. 2 (2008): 220–42.

146 *"I had not really intended initially"*: Baldomero Olivera, interview, Bishop Museum, Honolulu, Hawaii, June 5, 2015.

151 *what Toto calls the lightning-strike cabal*: Russell W. Teichert et al., "The molecular diversity of conoidean venom peptides and their targets: From basic research to therapeutic applications," in *Venoms to Drugs: Venom as a Source for the Development of Human Therapeutics*, ed. Glenn F. King, RSC Drug Discovery Series 42 (London: Royal Society of Chemistry, 2015), 163–203.

151 *putting whole schools of fish into an insulin coma*: Helena Safavi-Hemami et al., "Specialized insulin is used for chemical warfare by fish-hunting cone snails," *PNAS* 112, no. 6 (2015): 1743–48.

151 *distinct predatory and defensive venoms*: Sébastien Dutertre et al., "Evolution of separate predation- and defence-evoked venoms in carnivorous cone snails," *Nature Communications* 5 (2014): 3521.

151 *more than five hundred species in the genus* Conus: Thomas F. Duda Jr. and Alan J. Kohn, "Species-level phylogeography and evolutionary history of the hyperdiverse marine gastropod genus *Conus*," *Molecular Phylogenetics and Evolution* 34, no. 2 (2005): 257–72.

151 *ten thousand venomous marine snail species*: Teichert et al., 164; Baldomero M. Olivera et al., "Biodiversity of cone snails and other venomous marine gastropods: Evolutionary success through neuropharmacology," *Annual Review of Animal Biosciences* 2, no. 1 (2014): 487–513.

151 *a few hundred to several thousand different toxins*: Vincent Lavergne et al., "Optimized deep-targeted proteotranscriptomic profiling reveals unexplored *Conus* toxin diversity and novel cysteine frameworks," *PNAS* 112, no. 29 (2015): E3782–91.

152 *among the fastest-evolving DNA sequences*: Dan Chang and Thomas F. Duda Jr., "Extensive and continuous duplication facilitates rapid evolution and diversification of gene families," *Molecular Biology and Evolution* 28, no. 8 (2012): 2019–29.

152 *evolution as a change in the frequency of gene variations*: It's unclear exactly when it was so defined, but population geneticists have defined evolution as "change in allele frequencies in a population" since approximately the 1920s to the 1930s—see Marion Blute, "Is it time for an updated 'eco-evo-devo'definition of evolution by natural selection?" *Spontaneous Generations: A Journal for the History and Philosophy of Science* 2, no. 1 (2008): 1–5.

153 *an essential component of evolution*: Karen D. Crow and Günter P. Wagner, "What is the role of genome duplication in the evolution of complexity and diversity?" *Molecular Biology and Evolution* 23, no. 5 (2006): 887–92.

153 *on average, 1.13 times every million years*: Dan Chang and Thomas F. Duda Jr., 2023.

154 *The rate of non-synonymous substitutions in conotoxins*: Thomas F. Duda Jr. and Stephen R. Palumbi, "Molecular genetics of ecological diversification: Duplication and rapid evolution of toxin genes of the venomous gastropod *Conus*," *PNAS* 96, no. 12 (1999): 6820–23.

154 23 percent per million years: Chang and Duda, 2012.

155 *able to switch to fish prey*: Ai-Hua Jin et al., "δ-Conotoxin SuVIA suggests an evolutionary link between ancestral predator defence and the origin of fish-hunting behaviour in carnivorous cone snails," *Proceedings of the Royal Society B: Biological Sciences* 282 (2015): 20150817.

156 *don't so much target ion channels as make them*: Thomas C. Südhof, "α-Latrotoxin and its receptors: Neurexins and CIRL/latrophilins," *Annual Review of Neuroscience* 24 (2001): 933–62.

156 *Victims who experience systemic effects*: John Ashurst, Joe Sexton, and Matt Cook, "Approach and management of spider bites for the primary care physician," *Osteopathic Family Physician* 3, no. 4 (2011): 149–53.

156 *whose venom can cause seizures and comas*: M. Ismail, M. A. Abd-Elsalam, and M. S. Al-Ahaidib, "*Androctonus crassicauda* (Olivier), a dangerous and unduly neglected scorpion—I. Pharmacological and clinical studies," *Toxicon* 32, no. 12 (1994): 1599–1618.

156 *aptly named deathstalker scorpion*: Neil A. Castle and Peter N. Strong, "Identification of two toxins from scorpion (*Leiurus quinquestriatus*) venom which block distinct classes of calcium-activated potassium channel," *FEBS Letters* 209, no. 1 (1986): 117–21; Maria L. Garcia et al., "Purification and characterization of three inhibitors of voltage-dependent K^+ channels from *Leiurus quinquestriatus* var. *hebraeus* venom," *Biochemistry* 33, no. 22 (1994): 6834–39.

156 *reported to kill from 8 to 40 percent of human victims*: M. Ismail, "The scorpion envenoming syndrome," *Toxicon* 33, no. 7 (1995): 825–58.

156 *potent peptides to paralyze their intended prey*: C. Y. Lee, "Elapid neurotoxins and their mode of action," *Clinical Toxicology* 3, no. 3 (1970): 457–72.

157 *their folded shape is a core with three loops*: Carmel M. Barber, Geoffrey K. Isbister, and Wayne C. Hodgson, "Alpha neurotoxins," *Toxicon* 66 (2013): 47–58.

8. Mind Games

158 *"No human being had ever made me feel like that"*: Benjamin Alire Sáenz, *Last Night I Sang to the Monster* (El Paso, TX: Cinco Puntos Press, 2009), 26. The quote is about cocaine, but some say that the high of snake venom is strikingly similar.

159 *inject venom directly into subsections of its brain*: Ram Gal et al., "Sensory arsenal on the stinger of the parasitoid jewel wasp and its possible role in identifying cockroach brains," *PLoS ONE* 9, no. 2 (2014): e89683; Frederic Libersat and Ram Gal, "Wasp voodoo rituals, venom-cocktails, and the zombification of cockroach hosts," *Integrative and Comparative Biology* 54, no. 2 (2014): 129–42.

160 *in search of the missing brain regions*: Libersat and Gal, "Wasp voodoo rituals," 132.

160 *does not elicit the same hygienic urge*: Ibid., 133–34.

160 *the cause of this germophobic behavior*: A. Weisel-Eichler, G. Haspel, and F. Libersat, "Venom of a parasitoid wasp induces prolonged grooming in the cockroach," *Journal of Experimental Biology* 202, part 8 (1999): 957–64.

161 *floods of dopamine are triggered by pleasurable things*: Wolfram Schultz, "Dopamine signals for reward value and risk: Basic and recent data," *Behavioral and Brain Functions* 6 (2010): 24.

161 *we feel from illicit substances like cocaine*: Pradeep G. Bhide, "Dopamine, cocaine and the development of cerebral cortical cytoarchitecture: A review of current concepts," *Seminars in Cell and Developmental Biology* 20, no. 4 (2009): 395–402.

161 *the cockroach has lost all will to flee*: Ram Gal and Frederic Libersat, "On predatory wasps and zombie cockroaches: Investigations of free will and spontaneous behavior in insects," *Communicative and Integrative Biology* 3, no. 5 (2010): 458–61.

161 *the zombification wears off within a week*: Ram Gal and Frederic Libersat, "A parasitoid wasp manipulates the drive for walking of its cockroach prey," *Current Biology* 18, no. 12 (2008): 877–82.

161–62 *don't evoke a behavioral response*: Ibid.

162 *willingness to be buried and eaten alive*: Ram Gal and Frederic Libersat, "A wasp manipulates neuronal activity in the sub-esophageal ganglion to decrease the drive for walking in its cockroach prey," *PLoS ONE* 5, no. 4 (2010): e10019.

162 *the same chloride receptors, β-alanine and taurine*: Eugene L. Moore et al., "Parasitoid wasp sting: A cocktail of GABA, taurine, and β-alanine opens chloride channels for central synaptic block and transient paralysis of a cockroach host," *Journal of Neurobiology* 66, no. 8 (2006): 811–20.

163 *the stung cockroaches live longer*: Gal Haspel et al., "Parasitoid wasp affects metabolism of cockroach host to favor food preservation for its offspring," *Journal of Comparative Physiology A* 191, no. 6 (2005): 529–34.

163 *named for the soul-sucking guards*: Michael Ohl et al., "The soul-sucking wasp by popular acclaim—museum visitor participation in biodiversity discovery and taxonomy," *PLoS ONE* 9, no. 4 (2014): e95068.

164 *lay their eggs in spiders, caterpillars, and ants*: Ian D. Gauld, "Evolutionary patterns of host utilization by ichneumonoid parasitoids (Hymenoptera: Ichneumonidae and Braconidae)," *Biological Journal of the Linnean Society* 35, no. 4 (1988): 351–77; Jeremy A. Miller et al., "Spider hosts (Arachnida, Araneae) and wasp parasitoids (Insecta, Hymenoptera, Ichneumonidae, Ephialtini) matched using DNA barcodes," *Biodiversity Data Journal* 1 (2013): e992.

164 *to attach her eggs to caddisfly larvae*: J. M. Elliott, "The responses of the aquatic parasitoid *Agriotypus armatus* (Hymenoptera: Agriotypidae) to the spatial distribution and density of its caddis host *Silo pallipes* (Trichoptera: Goeridae)," *Journal of Animal Ecology* 52, no. 1 (1983): 315–30.

164 *nightmarish jaws of an ant lion*: H. Charles J. Godfray, *Parasitoids: Behavioral and Evolutionary Ecology* (Princeton, NJ: Princeton University Press, 1994), 290.

164 *lay eggs in the freshly pupated wasp larvae*: Jeffrey A. Harvey, Leontien M. A. Witjes, and Roel Wagenaar, "Development of hyperparasitoid wasp *Lysibia nana* (Hymenoptera: Ichneumonidae) in a multitrophic framework," *Environmental Entomology* 33, no. 5 (2004): 1488–96.

164 *defend pupating young wasps*: Amir H. Grosman et al., "Parasitoid increases survival of its pupae by inducing hosts to fight predators," *PLoS ONE* 3, no. 6 (2008): e2276.

164 *Another species' larva forces its spider host*: William G. Eberhard, "Spider manipulation by a wasp larva," *Nature* 406 (2000): 255–56.

165 *as other illicit drugs in India*: "V-Day drug: Youngsters get high on cobra venom," IBN Live, Feb 16, 2012, www.ibnlive.com/news/india/v-day-drug-youngsters-get-high-on-cobra-venom-447292.html.

165 *one liter fetching as much as 20 million rupees*: "Youth held with 'cobra venom' worth Rs 2 crore," *The Times of India*, February 6, 2014, http://timesofindia.indiatimes.com/city/lucknow/Youth-held-with-cobra-venom-worth-Rs-2-crore/articleshow/29921434.cms.

165 *teaming up with wildlife experts*: "'Cobra venom': Six accused in Forest custody, officials say they are only couriers," *The Peninsula*, June 30, 2014, http://thepeninsulaqatar.com/news/india/346809/cobra-venom-six-accused-in-forest-custody-officials-say-they-are-only-couriers.

165 *to crack down on the illegal sales*: Vijay Kautilya and Pravir Bhodka, "Snake venom—The new rage to get high!" *Journal of the Indian Society of Toxicology* 8, no. 1 (2012): 46–48.

166 *criminals caught with condoms full of snake venom*: "Drug Addicts Getting High with Snake Bites," Gulte.com, April 14, 2015, www.gulte.com /news/37793/Drug-Addicts-Getting-High-with-Snake-Bites.

166 *precious liquid, worth more than $15,000,000*: Rs. 100 crore = 100 x Rs. 10,000,000 = Rs. 1,000,000,000, converts to $15,628,310.00 USD (according to Google conversion rate, 2015); Rs. 100 crore from "'Cobra venom': Six accused," *The Peninsula*.

166 *and for violating protected-species laws*: Chandra S. Singh et al., "Species Identification from Dried Snake Venom," *Journal of Forensic Sciences* 57, no. 3 (2012): 826–28.

166 *graded as providing mild, moderate, or severe effects*: Subramanian Senthilkumaran et al., "Repeated snake bite for recreation: Mechanisms and implications," *International Journal of Critical Illness and Injury Science* 3, no. 3 (2013): 214–16.

166 *"the kick the other substances now lacked"*: Mohammad Zia Ul Haq Katshu et al., "Snake bite as a novel form of substance abuse: Personality profiles and cultural perspectives," *Substance Abuse* 32, no. 1 (2011): 43–46.

167 *"a sense of well-being, lethargy, and sleepiness"*: Katshu et al., 44.

167 *would have indulged daily, but the cost*: P. V. Pradhan et al., "Snake venom habituation in heroin (brown sugar) addiction: (Report of two cases)," *Journal of Postgraduate Medicine* 36, no. 4 (1990): 233–34.

167 *"a sense of well-being and happiness after each bite"*: Pradhan et al., 233–34.

167 *a high that he said lasted for several days*: "Teenager addicted to snake venom arrested in Kerala," *The Hindu*, August 18, 2014, www.thehindu.com /news/national/kerala/teenager-addicted-to-snake-venom-arrested-in -kerala/article6328195.ece.

167 *venom to help calm nerves and fight insomnia*: Senthilkumaran et al., 214–15.

167 *"quick effect" and an "extra kick"*: Pradhan et al.; T. K. Aich et al., "A comparative study on 136 opioid abusers in India and Nepal," *Journal of Psychiatrists' Association of Nepal* 2, no. 2 (2013): 11–17.

167 *so mild that they sleep right through them*: Utpal Jana and P. K. Maiti, "Dysphagia—an uncommon presentation of unnoticed snakebite," *Journal of the Indian Medical Association* 110, no. 9 (2012): 659–60.

167–168 *can be delayed for more than half an hour*: e.g., "Bitten by a Deadly Cobra," *Animal Planet*, www.animalplanet.com/tv-shows/fatal-attractions/videos /bitten-by-deadly-cobra/.

169 *"I did not wonder if it would ever end"*: Bryan Grieg Fry, *Venom Doc: The Edgiest, Darkest and Strangest Natural History Memoir Ever* (Sydney: Hachette Australia, 2015), 46–47.

170 *rejuvenated by very low doses of cobra venom*: Ludwin, "IAmA guy who's been injecting deadly snake venom."

170 *described feeling a "good, clean high"*: Phone interview, Anson Castelvecchi, August 3, 2015.

170 *"There is no death in the cup"*: Lucan, *The Civil War*, trans. James Duff, Book IX (London: William Heinemann, 1928).

171 *Brian Hanley, the founder of and chief scientist for Butterfly Sciences*: Brian Hanley, e-mail interview, August 3, 2015, http://bf-sci.com/?page_id=44.

171 *similar to the party drug γ-hydroxybutyric acid (or GHB)*: Hanley, e-mail interview.

171 *allowing it to free its head and bite his hand*: John Virata, "Kentucky Reptile Zoo Director Survives 9th Venomous Snake Bite in 38 Years," *Reptiles Magazine*, February 4, 2015, www.reptilesmagazine.com/Snakes/Information-News/Kentucky-Reptile-Zoo-Director-Survives-9th-Venomous-Snake-Bite-in-38-Years/.

171 *According to Jim, whose experiences*: Jim Harrison, phone interview, August 5, 2015.

172 *experiments in the early twentieth century*: David I. Macht, "Experimental and clinical study of cobra venom as an analgesic," *PNAS* 22, no. 1 (1936): 61–71.

172 *used by veterinarians as a horse analgesic*: D. De Klobusitzky, "Animal venoms in therapy," in *Venomous Animals and their Venoms*, vol. 3: *Venomous Invertebrates*, eds. Bücherl and Buckley (New York: Academic Press, 1971), 443–78; Ludovic Bailly-Chouriberry et al., "Identification of α-cobratoxin in equine plasma by LC-MS/MS for doping control," *Analytical Chemistry* 85, no. 10 (2013): 5219–25.

173 *to make neurons easier to trigger*: Benjamí Oller-Salvia, Meritxell Teixidó, and Ernest Giralt, "From venoms to BBB shuttles: Synthesis and blood–brain barrier transport assessment of apamin and a nontoxic analog," *Peptide Science* 100, no. 6 (2013): 675–86.

173 *apamin injection improves learning and cognitive performance*: O. Deschaux and J-C. Bizot, "Apamin produces selective improvements of learning in rats," *Neuroscience Letters* 386, no. 1 (2005): 5–8; F. J. Van der Staay et al., "Behavioral effects of apamin, a selective inhibitor of the SK_{Ca}-channel, in mice and rats," *Neuroscience and Biobehavioral Reviews* 23, no. 8 (1999): 1087–1110.

173 *or the lack of inhibitory signals*: Carl R. Lupica and Arthur C. Riegel, "Endocannabinoid release from midbrain dopamine neurons: A potential substrate for cannabinoid receptor antagonist treatment of addiction," *Neuropharmacology* 48, no. 8 (2005): 1105–16.

173 *why people feel a rush from bites*: Alexey Osipov and Yuri Utkin, "Effects of snake venom polypeptides on central nervous system," *Central Nervous System Agents in Medicinal Chemistry* 12, no. 4 (2012): 315–28.

173 *scientists can detect it in the mouse brains*: R. T. Gomes et al., "Comparison of the biodistribution of free or liposome-entrapped *Crotalus durissus terrificus* (South American rattlesnake) venom in mice," *Comparative Biochemistry and Physiology Part C: Toxicology and Pharmacology* 131, no. 3 (2002): 295–301.

173 *make their way into the central nervous system*: J. A. Alves da Silva, K. C. Oliveira, and M.A.P. Camillo, "Gyroxin increases blood-brain barrier permeability to Evans blue dye in mice," *Toxicon* 57, no. 1 (2011): 162–67.

174 *to induce effects from the outside*: Adriana C. Mancin et al., "The analgesic activity of crotamine, a neurotoxin from *Crotalus durissus terrificus* (South American rattlesnake) venom: A biochemical and pharmacological study," *Toxicon* 36, no. 12 (1998): 1927–37; Hui-Ling Zhang et al., "Opiate and acetylcholine-independent analgesic actions of crotoxin isolated from *Crotalus durissus terrificus* venom," *Toxicon* 48, no. 2 (2006): 175–82.

174 *direct action on the central nervous system*: For a review, see Osipov and Utkin.

174 *make their way past the brain's defenses*: Glenn F. King, "Venoms as a platform for human drugs: Translating toxins into therapeutics," *Expert Opinion on Biological Therapy* 11, no. 11 (2011): 1469–84.

174 *argument regarding* Cannabis *species*: Michael Pollan, *The Botany of Desire: A Plant's-Eye View of the World* (New York: Random House, 2001).

9. Lethal Lifesavers

177 *"from the lips of death the lessons of life"*: Felix Adler, *Life and Destiny: Or Thoughts from the Ethical Lectures of Felix Adler* (New York: McClure, Phillips and Company, 1903), 113.

177 *than all other deaths combined*: World Health Organization, "The top 10 causes of death," www.who.int/mediacentre/factsheets/fs310/en/index2.html.

177 *also dominate deaths under age fifty*: Ole F. Norheim et al., "Avoiding 40% of the premature deaths in each country, 2010–30: Review of national mortality trends to help quantify the UN Sustainable Development Goal for

health," *The Lancet* 385 (2015): 239–52, table 2; World Health Organization, "Burden: mortality, morbidity and risk factors," chapter 1 in *Global Status Report on Noncommunicable Diseases* 2010 (Geneva: WHO Press, 2011), 10; *The Global Burden of Disease, 2004 Update* (Geneva: WHO Press, 2008).

178 *Eng had never seen a Gila monster*: Andrew Pollack, "Lizard-Linked Therapy Has Roots in the Bronx," *The New York Times*, September 21, 2002, www.nytimes.com/2002/09/21/business/lizard-linked-therapy-has-roots-in-the-bronx.html.

178 *in* The Salt Lake Tribune *from 1898*: "The Horrible Gila Monster," *The Daily Tribune*, January 2, 1898, 15, www.newspapers.com/image/11120062/.

179 *"most to be dreaded of anything that crawls"*: "The Gila Monster and Its Deadly Bite," *The Milwaukee Journal*, November 1, 1898, 11.

179 *"the Gila monster to release its hold"*: "Terrors of the Gila Monster," *The San Francisco Call*, October 9, 1898, 23.

180 *at least one hundred times less potent than known killers*: Geeta Datta and Anthony T. Tu, "Structure and other chemical characterizations of gila toxin, a lethal toxin from lizard venom," *The Journal of Peptide Research* 50, no. 6 (1997): 443–50.

180 *a new peptide hormone that no one knew existed*: David Mendosa, *Losing Weight with Your Diabetes Medication: How Byetta and Other Drugs Can Help You Lose More Weight than You Ever Thought Possible* (Boston: Da Capo Press, 2008), chapters 5 and 6.

181 *told* The New York Times: Alex Berenson, "A Ray of Hope for Diabetics," *The New York Times*, March 2, 2006, www.nytimes.com/2006/03/02/business/02drug.html?_r=0.

181 *its first full year on the market*: FiercePharma, "Top 15 Drug Launch Superstars," October 2, 2013, www.fiercepharma.com/special-reports/top-15-drug-launch-superstars.

182 *the exenatide diabetes duo*: "AstraZeneca completes the acquisition of Bristol-Myers Squibb share of global diabetes alliance," AstraZeneca press release, February 3, 2014, www.astrazeneca.com/our-company/media-centre/press-releases/2014/astrazeneca-aquisition-bristol-myers-squibb-global-diabetes-alliance-03022014.html.

182 *preclinical testing of exendin-4 in the 1990s*: "Exendin-4: From lizard to laboratory . . . and beyond." NIH National Institute on Aging Newsroom, July 11, 2012, www.nia.nih.gov/newsroom/features/exendin-4-lizard-laboratory-and-beyond.

182 *began a human clinical trial*: "Exendin-4 in Alzheimer's Disease," www.nia.nih.gov/alzheimers/clinical-trials/exendin-4-alzheimers-disease.

182 *expected to rise to $2 trillion by 2030*: Martin Prince et al., *World Alzheimer Report 2015: The Global Economic Impact of Dementia* (London: Alzheimer's Disease International, 2015), www.alz.co.uk/research/world-report-2015.

183 *"I think the potential is still greater"*: Glenn King, interview, University of Queensland, Brisbane, Australia, November 28, 2014.

183 *having literally edited the book on it*: Glenn F. King, ed., *Venoms to Drugs: Venom as a Source for the Development of Human Therapeutics* (London: Royal Society of Chemistry, 2015).

184 *as a topical cure for baldness*: Markus Hellner et al., "Apitherapy: Usage and experience in German beekeepers," *Evidence-Based Complementary and Alternative Medicine* 5, no. 4 (2008): 475–79.

184 *used bee stings to treat gout*: Martin Grassberger et al., *Biotherapy—History, Principles and Practice: A Practical Guide to the Diagnosis and Treatment of Disease Using Living Organisms* (Dordrecht: Springer Netherlands, 2013), 78–80, http://link.springer.com/book/10.1007/978-94-007-6585-6.

184 *frequently employed snake venoms as therapeutics*: A. Gomes, "Snake Venom—An Anti Arthritis Natural Product," *Al Ameen Journal of Medical Sciences* 3, no. 3 (2010): 176.

184 *prevented blood loss from a life-threatening wound*: Adrienne Mayor, "The Uses of Snake Venom in Antiquity," *Wonders and Marvels*, November 2011, www.wondersandmarvels.com/2011/11/the-uses-of-snake-venom-in-antiquity.html.

185 *the use of cobra venom for pain*: John Henry Clarke, *A Dictionary of Practical Materia Medica* (London: The Homeopathic Publishing Company, 1902).

185 *followed up with experiments on people*: Robert N. Rutherford, "The use of cobra venom in the relief of intractable pain," *New England Journal of Medicine* 221, no. 11 (1939): 408–13.

185 *improved the symptoms of multiple sclerosis*: Ahmad G. Hegazi et al., "Novel therapeutic modality employing apitherapy for controlling of multiple sclerosis," *Journal of Clinical and Cellular Immunology* 6, no. 1 (2015): 299.

185 *snake venom to treat arthritis*: Antony Gomes et al., "Anti-arthritic activity of Indian monocellate cobra (*Naja kaouthia*) venom on adjuvant induced arthritis," *Toxicon* 55, no. 2 (2010): 670–73.

185 *Ellie told me that she became incapacitated*: Ellie Lobel, phone interviews, July 16, 2014, and January 23, 2015. Details also appeared in Christie Wilcox, "How a Bee Sting Saved My Life," *Mosaic*, March 24, 2015, http://mosaicscience.com/story/how-bee-sting-saved-my-life-poison-medicine.

186 *is a potent antibiotic*: Jean F. Fennell, William H. Shipman, and Leonard J. Cole, "Antibacterial action of a bee venom fraction (melittin) against a penicillin-resistant staphylococcus and other microorganisms," Research and Development Technical Report USNRDL-TR-67-101 (San Francisco: Naval Radiological Defense Lab, 1967).

186 *melittin has no trouble with them*: Lori L. Lubke and Claude F. Garon, "The antimicrobial agent melittin exhibits powerful in vitro inhibitory effects on the Lyme disease spirochete," *Clinical Infectious Diseases* 25, Supplement 1 (1997): S48–S51.

186 *worst symptoms in chronic Lyme sufferers*: Juliana Silva et al., "Pharmacological alternatives for the treatment of neurodegenerative disorders: Wasp and bee venoms and their components as new neuroactive tools," *Toxins* 7, no. 8 (2015): 3179–3209.

187 *Ken Winkel, the former head of the Australian Venom Research Unit*: Kenneth Winkel, interview, University of Melbourne, Melbourne, Australia, November 23, 2014.

188 *Sea anemone venom tackling autoimmune disorders*: "Kineta's ShK-186 shows encouraging early results as potential therapy for autoimmune eye diseases," press release, March 19, 2015, www.kinetabio.com/press_releases/Press Release20150319.pdf.

188 *Tarantula venom for muscular dystrophy*: Charlotte Hsu, "Good Venom," 2012, www.buffalo.edu/home/feature_story/good-venom.html.

188 *Centipede venom to cure unrelenting, excruciating pain*: Shilong Yang et al., "Discovery of a selective $Na_V1.7$ inhibitor from centipede venom with analgesic efficacy exceeding morphine in rodent pain models," *PNAS* 110, no. 43 (2013): 17534–39.

188 *cancer treatments lurking in the venoms of bees*: Nada Oršolić, "Bee venom in cancer therapy," *Cancer and Metastasis Reviews* 31, no. 1 (2012): 173–94.

188 *snakes, snails, scorpions*: Snakes: Vagish Kumar Laxman Shanbhag, "Applications of snake venoms in treatment of cancer," *Asian Pacific Journal of Tropical Biomedicine* 5, no. 4 (2015): 275–76; Snails: Shiva N. Kompella et al., "Alanine scan of α-conotoxin RegIIA reveals a selective α3β4 nicotinic acetylcholine receptor antagonist," *Journal of Biological Chemistry* 290, no. 2 (2015): 1039–48; Scorpions: "Scorpion venom has toxic effects against cancer cells," news release, Investigación y Desarollo, *AlphaGalileo*, May 27, 2015, www .alphagalileo.org/ViewItem.aspx?ItemId=153094 & CultureCode=en.

188 *and even mammals*: Quentin Casey, "Taming of the shrew's venom," *Financial Post*, July 2, 2012, http://business.financialpost.com/entrepreneur /taming-of-the-shrews-venom.

188 *making it one step closer to market*: Julie Fotheringham, "Targeting TRPV6 with Soricimed's novel SOR-C13 inhibits tumour growth in breast and ovarian cancer models," press release, *Market Watch*, May 6, 2015, www .marketwatch.com/story/targeting-trpv6-with-soricimeds-novel-sor -c13-inhibits-tumour-growth-in-breast-and-ovarian-cancer-models -2015-05-06.

189 *guide them during brain tumor surgery in children*: Study of BLZ-100 in Pediatric Subjects with CNS Tumors, Blaze Bioscience, Inc., https://clinicaltrials .gov/ct2/show/NCT02462629.

189 *bee venom can attack and kill human immunodeficiency virus*: David Fenardet et al., "A peptide derived from bee venom–secreted phospholipase A$_2$ inhibits replication of T-cell tropic HIV-1 strains via interaction with the CXCR4 chemokine receptor," *Molecular Pharmacology* 60, no. 2 (2001): 341–47.

189 *1.5 million deaths worldwide every year*: World Health Organization, Global Health Observatory Data, "Number of deaths due to HIV/AIDS," www .who.int/gho/hiv/epidemic_status/deaths/en/.

189 *compounds in snake venoms have shown activity against malaria*: Renaud Conde et al., "Scorpine, an anti-malaria and anti-bacterial agent purified from scorpion venom," *FEBS Letters* 471, no. 2 (2000): 165–68; Helge Zieler et al., "A snake venom phospholipase A$_2$ blocks malaria parasite development in the mosquito midgut by inhibiting ookinete association with the midgut surface," *Journal of Experimental Biology* 204, part 23 (2001): 4157–67.

189 *and reduce suffering in millions more*: World Health Organization, "10 Facts on Malaria," updated November 2015, www.who.int/features/factfiles /malaria/en/.

189 *that might be able to straighten that out*: Kenia P. Nunes et al., "Erectile function is improved in aged rats by PnTx2-6, a toxin from *Phoneutria nigriventer* spider venom," *Journal of Sexual Medicine* 9, no. 10 (2012): 2574–81.

189 *Bee venoms might be better than Botox*: Greg Ward, "Bee stings could be new Botox," BBC News, December 21, 2012, www.bbc.com/news/business -20807198.

189 *potential spermicide in black widow spider venom*: Antonio De La Jara, "Chile's Black Widow Spider May Yield Spermicide," Reuters, June 1, 2007, www .reuters.com/article/2007/06/01/us-chile-spider-idUSN0132580120070601.

190 *Rattlesnakes are gleefully rounded up*: Clark E. Adams et al., "Texas rattle-snake roundups: Implications of unregulated commercial use of wildlife," *Wildlife Society Bulletin* 22, no. 2 (1994): 324–30.

190–91 *to sell as pets, props, or parts*: K. Anna I. Nekaris et al., "Exploring cultural drivers for wildlife trade via an ethnoprimatological approach: A case study of slender and slow lorises (*Loris* and *Nycticebus*) in South and Southeast Asia," *American Journal of Primatology* 72, no. 10 (2010): 877–86.

191 *during life's 3-to-4-billion-year history*: Gerardo Ceballos et al., "Accelerated modern human–induced species losses: Entering the sixth mass extinction," *Science Advances* 1, no. 5 (2015): e1400253.

ACKNOWLEDGMENTS

I have to start by thanking my editor, Amanda Moon. Having spent the seven or so years before embarking on this challenge as an unedited blogger, I was completely unaware that good editors possess what I can only describe as supernatural powers. If you have enjoyed reading this book, it is because Amanda used her magic keyboard to transform my writing into page-turning prose. She and Scott Borchert, and everyone else at Scientific American / Farrar, Straus and Giroux, have been so patient and supportive, I could not have asked for a better publishing team. I also would be remiss if I didn't thank Eric Nelson, who suggested I write a book in the first place. Eric and Sue, and everyone else at the Susan Rabiner Literary Agency, have been a blessing to work with, and I am so grateful they had such faith in this first-time author.

No book like this gets written without the aid of many individuals and institutions—so many that I'm sure I'll forget some (please forgive me!), but I'll do my best to be complete. The venom community is rich with kind, generous people, many of whom I relied upon immensely to create this volume. Bryan Fry, Glenn King, Ken Winkel, Jim Harrison, and Toto Olivera: I deeply enjoyed picking your brains about venomous animals, and I could not have written this book without your insights and wisdom. Ellie Lobel: thank you for entrusting your incredible story to me. Thanks as well to the herpers who let me into their world, including Steve Ludwin, Anson Castelvecchi, Tim Friede, and Brian Hanley. And a very special thanks must go to those who helped me get up close to dangerous species: Chip Cochran, David Nelsen, Eric Gren, and Bill Hayes from Loma Linda University; Aaron Pomerantz, Jeff Cremer, and Frank Pichardo from Rainforest Expeditions; everyone at Komodo Dive Center and our local Rinca guide, Akbar; and Beck and the crew at Lone

Pine Koala Sanctuary (the thanks are for introducing me to these incredible creatures; the "very special" part is for making sure I walked away in one piece!).

Of course, I wouldn't be here in the first place if it weren't for two people whose mentorship has made me the scientist I am today. Brian Bowen: most Ph.D. advisors would have told me my blogging and social media were a waste of time and that I needed to focus more on my lab work, but you were never anything but supportive and encouraging of my extracurricular pursuits. Thank you for letting me be me—I'm sure it wasn't always easy. And Angel Yanagihara: I cannot imagine a more talented, meticulous, and passionate scientist to learn from. I know that my time as your postdoc has made me into the best scientist I can be, and I cannot express how grateful I am for your tutelage, friendship, and support.

Last, I have to thank my family and friends—my rocks who steadied me through the tumultuous book-writing process. Two of you in particular have gone above and beyond the call of duty. Kira Krend, you drill sergeant: thank you for cracking the whip and making sure I never gave up. And, of course, Jake Buehler, my sweet tea: I'll never forget our watching the bats emerge from a small island off Rinca from the top of a barely seaworthy boat after a long day of dragon hunting (*Superman!*). Thank you for being my traveling companion, my sounding board, and my nursemaid when I spent days staring at a screen, completely oblivious to the world outside my work, requiring reminders to eat and sleep. I could not have survived (perhaps quite literally) without you.

INDEX

Page numbers in *italics* refer to illustrations.

A NOTE ABOUT THE AUTHOR

Christie Wilcox, Ph.D., is a scientist and science writer based in Honolulu, Hawaii. Her writing has appeared in *Discover*, *The New York Times*, *Scientific American*, *Slate*, and *Popular Science*. Her website is www.christiewilcox.com. Follow her on Twitter at @NerdyChristie.